CRITICISM AND THE HISTORY OF SCIENCE
Kuhn's, Lakatos's and Feyerabend's Criticisms of Critical Rationalism

PHILOSOPHY OF HISTORY AND CULTURE

VOLUME 13

CRITICISM AND THE HISTORY OF SCIENCE

Kuhn's, Lakatos's and Feyerabend's Criticisms of Critical Rationalism

BY

GUNNAR ANDERSSON

E.J. BRILL
LEIDEN · NEW YORK · KÖLN
1994

The paper in this book meets the guidelines for permanence and durability of the Committee on Production Guidelines for Book Longevity of the Council on Library Resources.

Library of Congress Cataloging-in-Publication Data

Andersson, Gunnar.
[Kritik und Wissenschaftsgeschichte. English]
Criticism and the history of science : Kuhn's, Lakatos's, and Feyerabend's criticisms of critical rationalism / by Gunnar Andersson.
p. cm. — (Philosophy of history and culture, ISSN 0922-6001 ; v. 13)
Translation of: Kritik und Wissenschaftsgeschichte.
Includes bibliographical references and index.
ISBN 9004100504 (alk. paper)
1. Science—Philosophy. 2. Skepticism. 3. Rationalism. 4. Kuhn, Thomas S. 5. Lakatos, Imre. 6. Feyerabend, Paul K., 1924– . I. Title. II. Series.
Q175.A61813 1994
501—dc20 94-26308
CIP

Deutsche Bibliothek – CIP-Einheitsaufnahme

Andersson, Gunnar:
Criticism and the history of science : Kuhn's, Lakatos's and Feyerabend's criticisms of critical rationalism / by Gunnar Andersson. – Leiden ; New York ; Köln : Brill, 1994
(Philosophy of history and culture ; Vol. 13)
ISBN 90–04–10050–4
NE: GT

ISSN 0922-6001
ISBN 90 04 10050 4

Originally published in 1988 as
Kritik und Wissenschaftsgeschichte: Kuhns, Lakatos' und Feyerabends Kritik des Kritischen Rationalismus
by J.C.B. Mohr (Paul Siebeck), Tübingen.

PRINTED IN THE NETHERLANDS

TO LIDA

TABLE OF CONTENTS

PREFACE

When this book originally appeared in German, I was encouraged by some reviews to translate it into English. My analysis of Thomas Kuhn's, Imre Lakatos's and Paul Feyerabend's criticism of Karl Popper's falsificationist conception of science shows that their criticism is based on important methodological problems: the problem that observations and test statements are impregnated with theory, and the problem of how to test complex theoretical systems. In order to solve these problems I show how test statements can be critically discussed and whole theoretical systems tested empirically. My development of the falsificationist conception of science allows a deeper understanding of science and its history than Kuhn's, Lakatos's and Feyerabend's conceptions.

Umeå, April 1994 Gunnar Andersson

CHAPTER ONE

INTRODUCTION

A. THE FALSIFICATIONIST THEORY OF SCIENCE

In *The Logic of Scientific Discovery*, first published in German in 1934, Karl Popper advances a falsificationist philosophy of science. His point of departure is a criticism of the inductivist view that experience can show scientific hypotheses to be true or highly probable. According to Popper, scientific hypotheses are bold conjectures that experience can refute, but not justify. In order to test scientific theories empirically, you first derive prognoses about observable events from them. If a derived prognosis is false, then the tested theory is false. Such a refutation of a scientific theory presupposes only deductive logic and thus avoids the highly problematic inductivist assumptions. If the empirical test of a derived prognosis is positive and the predicted event occurs, then the tested theory is corroborated. However, it is impossible to show that a corroborated theory is true or highly probable. In order to understand science, you do not have to make any such inductivist assumptions. According to Popper (1959, § 79) even the most well corroborated theory is fallible and remains hypothetical.

Since David Hume (1739–40, book 1, part 3) the problem of induction has remained unsolved. It is doubtful whether it can be solved positively within an empiricist philosophy of science (Russell 1961, 637–47; Stegmüller 1971,13–74). On the other hand, falsifications presuppose only ordinary deductive logic, not any problematic inductivist principles. Therefore Popper maintains that inductivism only leads to unnecessary problems, that you do not need any inductive principles in order to understand the historical development of science or in order to advance a philosophy of science. Scientific theories are neither discovered nor tested inductively, but are hypothetical assumptions that can be tested severely through attempts to refute them. As long as such attempts fail, theories are corroborated and can be defended in a critical discussion.

Another basic problem in the philosophy of science, which Popper solves in a surprisingly simple way, is the problem of demarcation, the problem of how to distinguish scientific hypotheses from unscientific ones. Before Popper, empiricist philosophers of science suggested that a hypothesis is scientific if and only if it can be verified by experience. However, in his criticism of inductivism Popper shows that no general hypothesis is verifiable in this way. Hence the attempt to characterise scientific hypotheses as verifiable fails. Instead Popper suggests that a hypothesis is scientific if and only if it is falsifiable, that is, if and only if it can be refuted by experience. According to this solution to the problem of demarcation a hypothesis is unscientific if it is unfalsifiable and hence does not have any empirical consequences (Popper 1959, § 6; Popper 1979, 347–48).

B. The History of Science Challenges Falsificationism

In the fifties and sixties Norwood Russell Hanson, Stephen Toulmin, Thomas Kuhn, and Paul Feyerabend criticised the inductivist philosophy of science and advanced a "new philosophy of science." They extended Wittgenstein's criticism of logical empiricism into philosophy of science, a field that Wittgenstein had scarcely dealt with. They regarded the *logical* analysis used in logical empiricism as unfruitful and unsuitable for understanding the historical development of science. Instead they proposed a *historical* analysis of science (Radnitzky 1973, 56–71; Stegmüller 1973, 1–7). On the basis of such an analysis they criticised two basic positivist assumptions: the assumption of induction and the assumption of an infallible empirical basis of science. With historical case studies they argued that science neither develops inductively nor possesses any infallible empirical basis.

1. *Thomas S. Kuhn*

Thomas S. Kuhn criticises the inductivist view of science by showing that science does not develop cumulatively, as you would expect from an inductivist point of view, but that scientific revolutions regularly overthrow established scientific theories. So far Kuhn's interpretation of the history of science seems to fit a falsificationist view of science quite well, according to which old theories are replaced when they are refuted. However, Popper's descriptions of 'falsification' and 'refutation' strike Kuhn (1970a, 13) as extremely odd, since in science arguments are seldom as apodictic as in logic or mathematics. For this reason Kuhn does not speak of 'falsifications' and 'refutations', but of 'anomalies' and 'counter-examples'. He argues that scientific revolutions are not triggered off by anomalies alone, but that they must be preceded by a period of uncertainty in which the basic or paradigmatic theory fails to solve emerging problems. When such an instrumental and heuristic failure of the paradigm occurs, scientists lose faith in the paradigm, and a scientific crisis arises in which scientists regard anomalies as refuting counter-examples. Only then do scientists search for new theories and are shifts of paradigms possible. Thus a scientific crisis is a necessary precondition for a scientific revolution (Kuhn 1970b, chaps. 6–9).

According to Kuhn scientists do not try to test theories severely or to refute them. In "normal science" scientists accept a paradigm and use it as an instrument in order to solve problems. Kuhn avoids terms like 'problem' and 'problem solving' and uses instead 'puzzle' and 'puzzle solving'. In normal science there are always puzzles and anomalies, but they are not regarded as counterexamples and do not lead to new theories. Scientists reject an old theory first when they have suggested another and apparently better new theory. They suggest such theories first when there is a scientific crisis. Hence Kuhn (1970b, 77) writes: "No process yet disclosed by the historical study of scientific development at all resembles the methodological stereotype of falsification by direct comparison with nature." This seems to show that the falsificationist view of science is untenable. Kuhn thinks that scientific revolutions are rather rare events not caused by falsifications and that scientists do not test theories severely or try to refute them. These views are fundamental for the historical challenge of falsificationism.

Using examples from the history of science Kuhn (1970b, chap. 10) argues that observations and experiments are always interpreted with the help of some theory. After a scientific revolution many observations can be interpreted in quite a new way, and it even looks as if the scientists live in another world. In more cautious moments Kuhn says what scientists see when observing and experimenting depends on theoretical frameworks and paradigms. Hence Kuhn finds the positivist presupposition of an infallible and neutral empirical basis of science highly problematic. In any case he has not found any infallible and unrevisable test statements about observations and experiments in the history of science.

Kuhn thinks that the fallibility of test statements has consequences for falsificationism. He (Kuhn 1970a, 13–14) asks if falsifications are at all possible without sure test statements and maintains (Kuhn 1970b, 148–50) that competing theories are incommensurable since scientists with different paradigms accept different test statements. Is a critical discussion of theories, as Popper recommends, possible under such circumstances? Kuhn does not think so, but compares scientific revolutions to political revolutions, which violently change the whole social framework, and to religious conversions, which occur as sudden revelations. According to Kuhn (1970a, 21; 1970b, 92–94, 151–52) scientific change can only be explained psychologically and sociologically.

Kuhn's comparison of scientific revolutions to religious conversions is of course provocative. Logical empiricists and critical rationalists have agreed at least so far that science can be demarcated from religion and metaphysics. To many philosophers of science Kuhn's views appeared to be a dangerous form of relativism that threatened reason and science.

2. *Imre Lakatos*

Imre Lakatos tries to answer Kuhn's historical criticism of critical rationalism. He puts Popper's objectivism and *logic* of research against Kuhn's subjectivism and *psychology* of research (Lakatos 1970, 91–93). While Kuhn regards scientific revolutions as pseudoreligious conversions, Lakatos argues that they are reasonable and contribute to the growth of knowledge.

Like Kuhn, Lakatos considers the history of science and works with historical case studies. In order to decide whether falsificationism can counter the challenge from the history of science, Lakatos invents different versions of falsificationism.

In the first version, which Lakatos calls dogmatic falsificationism, it is assumed that experience can prove the truth of the empirical basis of science (Lakatos 1970, 95–103). This assumption is rejected in the second version, in methodological falsificationism, in which the empirical basis is laid down by conventional decisions (Lakatos 1970, 103–16). Lakatos considers two special versions of methodological falsificationism, one naive and one sophisticated. In the sophisticated version he tries to give "a new *rationale* of falsification and thereby rescue methodology and the idea of scientific *progress*" from Kuhn's relativism (Lakatos 1970, 116). In naive methodological falsificationism refuted theories are rejected. However, Lakatos does not think that this is in agreement with the history of science. Inspired by Kuhn, he thinks that the history of science shows that many refuted theories are not rejected. Lakatos (1968b, 163; 1970, 115) concludes paradoxically that the history of science falsifies naive methodological falsificationism. In sophisticated falsificationism a theory is rejected first when special criteria of elimination are met (Lakatos 1970, 116). These criteria are inspired by Popper's (1965, 240–42) requirements for the growth of knowledge. They can be interpreted as saying that a theory should be rejected when there is a better theory constituting objective growth of knowledge. Such objective criteria of progress certainly would overcome Kuhn's subjectivism. However, are there any objective criteria for the growth of knowledge? Can they be used in order to introduce new rules of elimination replacing logical falsifications?

Lakatos is not content with sophisticated falsificationism, but develops a methodology of scientific research programmes of his own, among other things in order to deal with the heuristic dimension of science. He rejects the view that the context of discovery is only psychologically interesting and tries to develop a rational heuristic method using the idea of research programmes (Lakatos 1970, 132–38). This attempt is inspired by Kuhn's ideas on the heuristic function of paradigms. Lakatos regards his methodology of scientific research programmes as an improvement of falsificationism. Many people share this opinion and regard the methodology of scientific research programmes as the most developed form of critical rationalism (Worrall 1978).

3. *Paul Feyerabend*

Paul Feyerabend denies that Lakatos succeeds in his attempts to save reason in science and argues that the history of science refutes not only falsificationism, but also the methodology of scientific research programmes. Feyerabend does not think that Lakatos's criteria for elimination of research programmes are rational. Firstly, new and young research programmes often have to struggle with considerable difficulties before the necessary auxiliary hypotheses and disciplines are developed. Secondly, you can never rule out the possibility of an unsuccessful and "degenerative" research programme developing and becoming successful and progressive. Therefore Feyerabend maintains that Lakatos does not succeed in saving science from the assault of historical analysis (Feyerabend 1976; Feyerabend 1978, 315).

Originally, Feyerabend called his own position epistemological anarchism, perhaps in order to take precautions against the accusation of being an epistemological nihilist (Stegmüller 1979, 262–63). Epistemological anarchism is in reality a type of historical relativism, in which Feyerabend draws relativist conclusions from the history of science more clearly than Kuhn did. Feyerabend's position is mainly of interest as a criticism of other positions. The historical case studies he uses in order to support his criticism also provide interesting material.

The thesis of incommensurability belongs to the positive content of epistemological anarchism. This thesis is also defended by Kuhn. Like Kuhn, Feyerabend thinks that this thesis is a consequence of the theory-dependence of experience (Feyerabend 1988, ch. 16). He does not think that critical discussion is sufficient in order to introduce new theories; rather, you have to behave in an anarchic way and use propaganda and persuasion. According to Feyerabend there is a rhetorical and propagandistic element in every scientific revolution. The only universally valid methodological rule is: anything goes (Feyerabend 1988, 19). This sounds peculiarly familiar. Anything is allowed if God is dead, the nihilists earlier said.

Thus Feyerabend's historical analysis of science ends in scepticism, a not quite unusual fate for students of history. He tries to meet this fate with composure. In his "plea for hedonism" he argues that you can nevertheless be quite happy, that truth is just an old woman (Feyerabend 1970, 229). In this way Feyerabend tries to give historical relativism, usually an expression of tired scepticism, more vivid and original traits.

C. Is Falsificationism Tenable?

1. *Some Reactions to the Criticism of Falsificationism*

Does the history of science show that falsificationism is untenable? Wolfgang Stegmüller thinks so. According to him historical arguments show that both logical empiricism and critical rationalism are untenable (Stegmüller 1973, 3–10). Therefore Stegmüller searches for new ways in philosophy of science and finds them in an instrumentalist conception of scientific theories, according to which theories are "non-statements," that is formal structures with intended applications. In the thirties Moritz Schlick regarded theories in a similar way. Inspired by Wittgenstein, Schlick maintained that universal statements are no genuine statements at all, but only apparent statements and instructions for the formation of statements (Popper 1979, chaps. 7, 9). Stegmüller hopes that his non-statement view of scientific theories will explain the historical development of science and the "dynamics of theories" better than the static analyses of logical empiricism did.

Elisabeth Ströker thinks that the study of the history of science can liberate critical rationalism from naive ideas of reality and truth. She thinks that the belief in an objective absolute truth of scientific knowledge is groundless if a historical process determines what scientists regard as real (Ströker 1976,60).

Lothar Schäfer regards Lakatos's methodology as an "enlightened falsificationism" and as a convincing and interesting methodology, but nevertheless thinks that Popper's arguments against Lakatos are valid. Therefore he pleads for methodological pluralism (Schäfer 1974, 223–24).

Kurt Hübner (1979, 396) regards the historical challenge of methodology as a symptom of a crisis of "scientific reason" today, even if most people are not aware of it. He thinks that critical rationalists are unhistorical and have a naive theory of truth. Hübner's own position is rather similar to Feyerabend's. Both maintain that the scientific and mythical world views are incommensurable. Hübner thinks that we can compare and understand them as alternatives, but that we do not have any superior standard for evaluating them.

2. *Basic Problems and Ideas*

What are the consequences of the historical challenge of falsificationism? I will argue that this challenge depends on methodological problems. In order to be able to discuss these problems, I will investigate the logical relations between explanations, predictions and falsifications, and deal with the problem of testing whole theoretical systems in chapter two below.

According to Stegmüller (1979, 2: 265–69) it is a basic presupposition of falsificationism that isolated hypotheses are tested (isolationism). In the chapter two I will show that falsificationism can deal very well with the testing and falsification of theoretical systems (holism). The methodological problems of such tests have been discussed by Duhem (1974, part 2, chap. 6) and Quine (1961, chap. 11). Much of the historical criticism of falsificationism has its origin in these problems.

In the following three chapters I will analyse the historical criticism of falsificationism in order to find out the most important arguments in the works of Kuhn, Lakatos and Feyerabend. As far as it is useful for this analysis, I will deal with some of their historical case studies. These historical investigations only have the function of making the analysis of their criticism clearer. I will especially try to find out in exactly what ways their studies of the history of science challenge falsificationism. Following Popper's (1974, 977) advice, I will try to present their criticism in its sharpest and best form, if possible still more convincingly than the critics themselves.

My analysis shows that the historical criticism of falsificationism is based on methodological problems. Of greatest importance and interest in this criticism are not the historical facts themselves, but the methodological problems they illustrate.

I will show that Kuhn's, Lakatos's and Feyerabend's criticism of falsificationism is based on two methodological problems: the problem of the theory-dependence of experience and the problem of the systematic or holistic character of empirical testing. Already conventionalists like Duhem (1974, chap. 4) and Le Roy (1899–1900) have discussed these problems. Kuhn and Feyerabend continue these discussions and use arguments similar to those used by the conventionalists against the positivist philosophy of science. Does Kuhn's and Feyerabend's neoconventionalist criticism using examples from the history of science hit critical rationalism? This is the basic problem of this book.

Popper discusses the theory-dependence of experience as the problem of the empirical basis. He (Popper 1959, chap. 5) criticises the positivist assumption that this "basis" is infallible and independent of theory. So far Popper agrees to the conventionalist criticism of positivism. However, he stresses that it is not enough to maintain that test statements are fallible. If you do not want to throw empiricism overboard, you have to solve the methodological problems arising from the fallibility of test statements (Popper 1959, § 26). Has Popper succeeded in solving these problems? Can theory-dependent and fallible test statements always be critically discussed, as Popper suggests?

Popper's solution to the problem of the empirical "basis" is also contested among philosophers of science close to critical rationalism (Agassi 1975, chap. 5; Bartley 1982, 162–68; Koertge 1978; Watkins 1984). Popper (1973, chap. 5, addendum 1968) maintains that his solution has partly been misunderstood. Does a proper understanding of it solve the methodological problems of test statements discussed by Kuhn, Lakatos, and Feyerabend?

In the chapter six I will deal with the problem of the empirical "basis." The main thesis of this chapter is that a critical discussion of fallible test statements is always possible. This thesis is not generally accepted. Kuhn and Feyerabend find Popper's solution to the problem of the empirical "basis" unsatisfactory and interpret it as positivist. Lakatos, on the other hand, interprets the same solution as conventionalist and tries to improve on it in his discussion of sophisticated falsificationism. I will argue that Popper suggests a solution to the problem of the empirical "basis" that is neither positivist nor conventionalist, but criticist. The point of departure of Popper's arguments is the intersubjective testability of test statements. In the chapter six I will show that it is possible to discuss fallible test statements critically and that the methodological problems arising from the theory-dependence of experience can be solved through such a discussion. Although I will concentrate on the methodological aspects of the problem, I think that it is epistemologically and philosophically relevant and interesting that the positivist assumption of an infallible empirical basis can be rejected without lapsing into relativism or conventionalism.

In the chapter seven I will deal with the methodological problems arising when whole theoretical systems are tested empirically. When such a system is falsified, several strategies are possible. You can try to suggest an entirely new system, or you can try to modify the refuted system more or less extensively. The last strategy is often regarded as methodologically objectionable, as an attempt to save refuted theories from criticism. Important parts of the historical criticism of falsificationism rest on the assumptions that critical rationalists do not allow modifications of falsified theories, but rather recommend a radical and revolutionary strategy of research, according to which refuted theories always have to be entirely rejected. Kuhn and Lakatos argue that science does not develop according to such a strategy of permanent revolution in science, but that small modifications of refuted theories are important in science. Does this mean that falsifications are not taken seriously, as Kuhn, Lakatos, and Feyerabend think? I do not think so. In the chapter seven I will argue that not only a revolutionary, but also a cautious and conservative strategy of research ought to be allowed in a falsificationist methodology. There are no logical or methodological reasons for the recommendation always to reject falsified theories. Modifications of a refuted theory show that a falsification is taken seriously, that scientists try to learn from experience. Therefore, some aspects of what Kuhn calls normal science do not have to be understood as dogmatic or as dangerous for the growth of knowledge, but as cautious and piecemeal attempts to learn from experience.

Some aspects of Lakatos's methodology of scientific research programmes and Feyerabend's (1970, 203) principle of tenacity can also be regarded in this way. If central parts of a research programme are kept unchanged, you do not have to explain this by assuming, as Lakatos does, that they have been made unfalsifiable by a methodological decision. You can regard it as an undogmatic attempt to solve problems by modifying the peripheral parts of the research programme, as an attempt to follow a cautious and conservative research strategy.

Auxiliary hypotheses are often necessary in order to test theories and hypotheses empirically. The Copernican theory was a rather complex theory. Nevertheless, auxiliary hypotheses were necessary in order to derive some important empirical consequences from it, as Feyerabend (1988, chaps. 6–10) has shown. For example, Galileo assumed that the telescope is a reliable instrument, an assumption that was problematic in the beginning. Tycho Brahe and other followers of Aristotelian physics thought that simple experiments could refute the Copernican hypothesis of the diurnal motion of the earth. Galileo argued that these falsifications depend on problematic auxiliary hypotheses and suggested new auxiliary hypotheses that brought the Copernican theory into agreement with experience.

The use of such new auxiliary hypotheses aggravates the problem of theoretical holism. Feyerabend argues that the auxiliary hypotheses suggested by Galileo in his defence of the Copernican theory were *ad hoc* and objectionable attempts to evade criticism. In the chapter seven I will argue that this is not the case, that Galileo's auxiliary hypotheses were neither *ad hoc* nor attempts to evade criticism, but necessary in order to test the astronomical theories and to discuss them critically. Feyerabend's attempt to make Galileo a patron of epistemological anarchism does not succeed.

In the methodological discussion it has remained unclear what hypotheses are *ad hoc* and why they are methodologically objectionable. Feyerabend's criticism of falsificationism depends partly on the unclear distinction between objectionable *ad hoc* hypotheses and useful auxiliary hypotheses. In the chapter seven I will clear up this distinction.

This distinction is also important for a discussion of Lakatos's ideas. He (Lakatos 1970, 100–103, 184–86) repeats the conventionalist objection that suitable auxiliary or *ad hoc* hypotheses can immunise any theory against criticism. It is not so easy to find good examples of *ad hoc* hypotheses in the history of science, probably because scientists have an instinctive aversion against such hypotheses and because such hypotheses are not recorded by the historians of science. However, some of Lakatos's historical thought experiments are rich mines for philosophers of science searching for examples of *ad hoc* hypotheses.

In the chapter seven I will argue that the possibility of introducing *ad hoc* hypotheses does not threaten falsificationism in a serious way. Popper (1959, § 22) requires that falsifications should be reproducible. To immunise a theory against reproducible falsifications with *ad hoc* hypotheses is not as easy as many philosophers of science have thought.

I will show that the historical criticism of falsificationism is methodological. With historical case studies Kuhn, Lakatos, and Feyerabend discuss methodological problems of falsificationism. Especially important are two main types of problems, which recur throughout their discussions. The first type of problem has to do with test statements and their dependence on theories and can be called the problem of test statements. The second type of problem depends on the holistic or systematic character of empirical testing. To this type belong the problem of how to judge modifications of falsified theories, the problem of how to evaluate different research strategies, and the problem of distinguishing between useful auxiliary hypotheses and objectionable *ad hoc* hypotheses.

The two types of methodological problems basic for the historical criticism of falsificationism can be solved in a criticist way. The solutions to these problems offered by falsificationism are superior to the solutions offered by Kuhn, Lakatos, and Feyerabend and allow a deeper understanding of the historical development of science. The history of science is a history of a critical discussion of bold conjectures proposed in order to explain observations and experiments. The falsificationist methodology shows how such a critical discussion is possible.

CHAPTER TWO

LOGICAL ASPECTS OF FALSIFICATION

In this chapter I will investigate the relations between falsifications on the one hand and explanations and predictions on the other. In order to be able to specify the logical structure of falsifications, I will deal with the logical structure of explanations and predictions first.

A. EXPLANATION

A basic aim of science is to find satisfactory explanations (Popper 1972, § 15). Here I will not discuss whether science has the still more basic aim to search for truth. In order to deal with many methodological problems, it is enough to assume that science has the comparatively unproblematic aim to find satisfactory explanations.

Prescientific myths and cosmologies are also attempts to explain events. Science continues these attempts, but partly in a new way. The decisive step in the transition from myth to science is the step from uncritical acceptance of explaining hypotheses as parts of an unquestioned tradition to critical discussion of them. Presocratic Greek philosophers were first to take this step and establish a critical tradition from which European philosophy and science emerged (Popper 1965, chap. 5).

In order to discuss scientific explanations, I will discuss a simple example. The point of departure of all research is some particular problem. Let us assume that we want to explain why a thread breaks. A possible answer to this question is to point out that all threads have a characterising tensile strength, that the tensile strength of this special thread is 1 *lb.*, and that a weight of 2 *lbs.* is put on it. This is a possible explanation of the breaking of the thread. It has the following logical structure (Popper 1959, § 12):

For every thread of a given structure S (determined by its material, thickness, etc.) there is a characteristic weight w, such that the thread will break if any weight exceeding w is suspended from it. (Hypothesis H_1)
For every thread of the structure S_1, the characteristic weight w_1 equals 1 *lb*. (Hypothesis H_2)
This is a thread of structure S_1. (Initial condition C_1)
The weight put on this thread is equal to 2 *lbs*. (Initial condition C_2)

This thread breaks. (Prognosis P)

This explanation is a deductive derivation of a description of the event to be explained (*explanandum*) from explaining propositions (*explanans*) consisting of universal hypotheses and initial conditions. The given explanation can be summarised in the following way:

(1) $H_1, H_2, C_1, C_2 \vdash P$

When we try to explain an event, e.g. the breaking of a thread, we try to derive a prediction of the event from universal hypotheses and specific initial conditions. The term 'prediction' is here used not only for statements about the future, but also for given statements that we want to explain (*explananda*) and even for statements about the past (retrodictions) (Popper 1959, 60, fn. *2). Explanations of this kind consisting of the deduction of a prognosis from general hypotheses I will call hypothetico-deductive explanations. They have the following general logical form (Hempel 1965, 249, 336; Stegmüller 1974, 86):

(2) $H_1, \ldots, H_r, C_1, \ldots, C_n \vdash P$

or

$$\frac{\begin{array}{c} H_1, \ldots, H_r \\ C_1, \ldots, C_n \end{array}}{P} \qquad \begin{array}{r} \text{(Universal hypotheses)} \\ \text{(Specific initial conditions)} \\ \text{(Prognosis)} \end{array}$$

Popper calls these explanations causal, but he does not assume any causal principle, neither that all events can be causally explained nor that the world is governed by deterministic laws. Such assumptions would be metaphysical. Instead he proposes a methodological rule, which can be understood as the methodological version of the metaphysical causal principle. "It is the simple rule that we are not to abandon the search for universal laws and for a coherent theoretical system, nor ever give up our attempts to explain causally any kind of event we can describe" (Popper 1959, 61).

B. PREDICTION

In the above example we cannot only explain why a thread breaks, but we can also predict that a thread will break under certain conditions. It depends upon our interests whether we use hypotheses in order to explain or in order to predict events. When we explain, we want to know why an event occurred. The event is given and the hypotheses and conditions explaining it are searched for. When we predict, we want to know what will happen in the future. Some hypotheses are given and a prognosis is searched for. Thus the difference between explanation and prediction is not logical, but pragmatic (Popper 1960, 133). Stegmüller (1974, 153) calls this the thesis of the structural similarity between explanation and prediction. According to this thesis explanation and prediction have the same logical form. It is, however, true only for explanation and *unconditional prediction*, that is for the derivation of unconditional prognoses of the type 'this thread breaks'. An example of an unconditional prediction is:

> For every thread of a given structure S (determined by its material, thickness, etc.) there is a characteristic weight w, such that the thread will break if any weight exceeding w is suspended from it. (H_1)
> For every thread of the structure S_1, the characteristic weight w_1 equals 1 *lb.* (H_2)
> This is a thread of structure S_1. (C_1)
> The weight that will be put on this thread is equal to 2 *lbs.* (C_2)
> ___
> This thread will break. (P)

This prediction has the same logical form as the explanation discussed above. For the logical form of the derivation it is unimportant whether the initial conditions and the prognosis describe future or past events. This difference is merely pragmatic. Since the prognosis in the prediction above is unconditional, the thesis of the structural similarity between explanation and prediction is true in this case.

Conditional prognoses are often interesting in science and technology, as for example: 'If this thread has the structure S_1 and a weight equal to 2 *lbs.* is put on it, then it will break.' The derivation of this conditional prognosis has the following logical form:

> For every thread of a given structure S (determined by its material, thickness, etc.) there is a characteristic weight w, such that the thread will break if any weight exceeding w is suspended from it. (H_1)
> For every thread of the structure S_1, the characteristic weight w_1 equals 1 *lb.* (H_2)
> ___
> If this is a thread of structure S_1 and the weight that will be put on it is equal to 2 *lbs.*, then this thread will break. ($(C_1 \wedge C_2) \rightarrow P$)

or

(3) $H_1, H_2, \vdash (C_1, \wedge C_2) \rightarrow P$

Not all conditions have to appear in the conclusion, that is to say in the prognosis. It is also possible that some belong to the premises, as in the following derivation:

(4) $H_1, H_2, C_1 \vdash C_2 \rightarrow P$

In this case the conditional prognosis 'If this thread is loaded with 2 *lbs.*, then it will break' is derived. The first initial condition, 'This thread has the structure S_1', belongs to the premises.

The general logical form for conditional predictions is:

$$\frac{\begin{array}{c} H_1, \ldots, H_r \\ C_1, \ldots, C_k \end{array}}{(C_{k+1} \wedge \ldots \wedge C_n) \rightarrow P} \qquad \begin{array}{r} \text{Universal hypotheses} \\ \text{Initial conditions } (0 \leq k \leq n) \\ \text{Conditional Prognosis} \end{array}$$

or

(5) $H_1, \ldots, H_r, C_1, \ldots, C_k \vdash (C_{k+1} \wedge \ldots \wedge C_n) \rightarrow P \ (0 \leq k \leq n)$

The derivations of conditional and unconditional prognoses have different logical forms. Thus they are structurally different. However, if you can derive an unconditional prognosis, you can also derive a conditional prognosis and vice versa. Thus the derivations of conditional and unconditional prognoses are metalogically equivalent. It depends only on the problem situation, that is on pragmatic circumstances, whether the initial conditions appear as premises or in the conclusion.

The metalogical equivalence between conditional and unconditional predictions is expressed in the following theorem:

(6) $H_1, \ldots, H_r, C_1, \ldots, C_n \vdash P$
if and only if
$H_1, \ldots, H_r, C_1, \ldots, C_k \vdash (C_{k+1} \wedge \ldots \wedge C_n) \rightarrow P \quad (0 \leq k \leq n)$

This theorem is proved in the appendix below as theorem 1 (b). It is important that the theorem states an equivalence between different valid derivations and thus is a metalogical theorem.

This metalogical equivalence can be illustrated in the following way. Above we derived the unconditional prognosis 'This thread will break' from a set of premises including the initial condition 'This thread is loaded with 2 *lbs.*' Then we can derive the conditional prognosis 'If this thread is loaded with 2 *lbs.*, it will break' according to Tarski's deduction theorem (Kleene 1967, 39–41, theorem 11) or to what in elementary logic is called the principle of conditional proof. That is:

(7) If $H_1, H_2, C_1, C_2 \vdash P$, then
$H_1, H_2, C_1 \vdash C_2 \rightarrow P$

The inversion of the metalogical implication (6) can also be proved with the inversion of Tarski's deduction theorem (Kleene 1967, 38–39, theorem 10):

(8) If $H_1, H_2, C_1 \vdash C_2 \rightarrow P$, then
$H_1, H_2, C_1, C_2 \vdash P$

These considerations show that we can always choose what type of prognosis to derive. It depends on our problem situation whether a conditional or unconditional prognosis is interesting. Thus we can choose if we want to derive the unconditional prognosis 'This thread will break' or a conditional prognosis, as for example 'If this thread is loaded with 2 *lbs.*, then it will break'.

Popper distinguishes between two different types of predictions, between prophecies and technological predictions. A prophecy tells us about "an event which we can do nothing to prevent," as for example the prognosis that a typhoon will come. A technological prediction, on the other hand, is constructive "intimating the steps open to us *if* we want to achieve certain results," as for example the prediction that a shelter has to be built in a certain way in order to stand up to a typhoon (Popper 1960, 43). What Popper calls a prophecy is an unconditional prediction with initial conditions that cannot be controlled technologically, whereas what he calls a technological prediction is a conditional prediction with initial conditions that can be controlled technologically.

A conditional prognosis states that a certain event will occur if certain initial conditions are satisfied. Such prognoses, having the form of if-then-statements, are not only technologically interesting but can also be used in order to test scientific theories empirically. For this reason Hempel (1966, 7) calls them *test implications*. Hence conditional predictions are constructive also in the sense that they tell us how to test theories. If a conditional prognosis has conditions that can be controlled technologically, then an experimental test is possible; otherwise the testing scientist has to wait for or search out cases in nature where the stated conditions are realised (Hempel 1966, 20).

When we try to explain an event, we try to derive an unconditional prognosis describing the event to be explained. When we predict an event, it depends on the circumstances whether a conditional or an unconditional prediction is more interesting. In astronomy unconditional prognoses are usually derived because the initial conditions for astronomical events, as for example eclipses or the positions of planets, cannot be influenced by us. In many other sciences, however, the initial conditions can be influenced, and thus conditional predictions are often scientifically and technologically more interesting. For example, the conditional prognosis 'If this thread has the structure S_1 and is loaded with 2 *lbs.*, then it breaks' can be used technologically to bring a thread to break. However, it is also an instruction suggesting how a theoretical system consisting of two hypotheses (cf. (3) above) can be tested experimentally. Thus the conditional prognosis can be understood both as a technological prognosis and as a test implication in this case. There is a close connection between experimental test and technological application. This connection is the basis for the modern justification of science, according to which knowledge is power that can make man master and possessor of nature (Bacon 1620; Descartes 1637). However, if our knowledge is fallible and incomplete, as Popper argues, then our power over nature will remain limited, and practical applications of science will be accompanied by unintended consequences (Popper 1960, § 21). Furthermore, scientific hypotheses also show what cannot be done and hence also show us the limits of man's power.

C. FALSIFICATION

1. *Falsification with Negations of Prognoses*

In *The Logic of Scientific Discovery* Popper deals with two different types of falsifications. In § 18 he justifies falsification with the modus tollens of classical logic: if we derive a prognosis from a theoretical system and observations show that the prognosis is false, then the theoretical system is false. Popper stresses that "by means of this mode of inference we falsify *the whole system* (the theory as well as the initial conditions) which was required for the deduction of the statement *p*. ... Thus it cannot be asserted of any one statement of the system that it is, or is not, specifically upset by the falsification" (Popper 1959, 76).

In the example above the prognosis 'This thread breaks' was derived. If we test this prognosis and observe that it is false, then we can assert the negation of our original prognosis, namely 'This thread does not break'. What follows from this negation? In order to derive the original prognosis we used two general hypotheses and two initial conditions as premises (cf. (1) above). The conjunction of these premises is false if the conclusion is false. This falsification hits a theoretical system consisting of general hypotheses and initial conditions:

(9) $\neg P \vdash \neg(H_1 \wedge H_2 \wedge C_1 \wedge C_2)$

Why is this inference valid? Popper's original justification in *The Logic of Scientific Discovery* using the modus tollens of classical logic is not quite satisfactory, as he later remarks (Popper 1959, 76, fn. *1). In his original justification Popper does not sufficiently distinguish metalogical statements about deducibility from logical if-then-statements. Although Popper's justification is unsatisfactory, the derivation (9) is nevertheless valid. How can it be justified correctly? The problem is to show that if a conclusion of a derivation is false, then the conjunction of the premises must be false. In our example we must show that the following metalogical implication is valid:

(10) If $H_1, H_2, C_1, C_2 \vdash P$,
then $\neg P \vdash \neg(H_1 \wedge H_2 \wedge C_1 \wedge C_2)$

Later, Popper has given a general metalogical justification for the validity of inferences like (10). He writes that deductive logic is a theory of how to transfer truth from the premises to the conclusion: if a derivation is valid and all premises true, then the conclusion *must* be true. It is impossible that all the premises in a valid derivation are true, if the conclusion is false. Thus deductive logic is not only a theory of the transfer of truth from the premises to the conclusion, but also a theory of the retransfer of falsity from the conclusion to at least one of the premises. In this way deductive logic becomes a theory of rational criticism (Popper 1970a, 116).

In the appendix below the following generalisation of (10) will be proved (as theorem 2b):

(11) $H_1, \ldots, H_r, C_1, \ldots, C_n \vdash P$
if and only if
$\neg P \vdash \neg(H_1 \wedge \ldots \wedge H_r \wedge C_1 \wedge \ldots \wedge C_n)$

This theorem shows that the first type of falsification dealt with by Popper is valid.

The equivalence (11) is methodologically interesting. If an unconditional prognosis can be derived from a theoretical system consisting of hypotheses and initial conditions and if the prognosis can be tested by observation, then the whole theoretical system can be empirically tested. If the predicted event does not occur, the theoretical system is falsified; if the event occurs, the system is corroborated.

This shows that a falsificationist methodology must not presuppose that only isolated hypotheses can be empirically tested. Whole theoretical systems can be tested, systems consisting of all the hypotheses and initial conditions necessary in order to derive the tested prognosis. Of course this does not rule out that in special cases a single hypothesis (together with one or more appropriate initial conditions) can be empirically tested. This is only a special case. A falsificationist methodology does not have to presuppose theoretical "isolationism," as Stegmüller (1979, 2: 265) suggests, but is very well compatible with a theoretical "holism."

The equivalence (11) also shows the close connection between falsifiability and explanatory power. A theoretical system that can explain and predict observable events is also falsifiable. If a predicted event does not occur, the theoretical system is false. On the other hand, if a theoretical system is falsifiable, then it can also predict and explain empirical events. Thus unfalsifiable theoretical systems are systems that cannot explain and predict observable events, that is they do not have any explanatory power.

2. *Falsification with Basic Statements*

In *The Logic of Scientific Discovery* Popper deals with a second type of falsifying argument: falsification of a general hypothesis with a basic statement. He determines the logical form of a basic statement in such a way that a basic statement can contradict a universal statement (Popper 1959, § 28). Thus falsifications of this type can refute a single universal hypothesis, whereas the type of falsification discussed earlier refuted a theoretical system consisting of general hypotheses and initial conditions.

An example of a falsifying inference of the second type is the following (Popper 1983, xx):

> On the 16th of May, 1934, a black swan stood between 10 and 11 o'-clock in the morning in front of the statue of Empress Elizabeth in the Volksgarten in Vienna.
> ___
> Not all swans are white.

In order to show that this type of falsification is valid, Popper uses a logical equivalency between special types of universal and existential statements. For example, 'Not all swans are white' is a negation of a universal statement and is logically equivalent to the existential statement 'There is a swan which is not white'. In this existential statement only universal names, like 'swan' and 'white', occur, and for this reason Popper calls it a strictly or purely existential statement (Popper 1959, § 15 with another example). In the falsifying basic statement there are not only universal names, but also individual ones telling us where and when there was a swan that is not white, in what special space-time-region: before the statue of empress Elisabeth in the Volksgarten in Vienna between 10 and 11 a.m. Popper calls such statements singular and requires that all basic statements shall have the form of singular existential statements. According to him a basic statement shall have the logical form: in a space-time region *k* there is a *P* (Popper 1959, § 28).

The basic statement 'In the space-time region *k* there is a non-white swan' has logical consequence 'There is a non-white swan'; this statement is logically equivalent to the statement 'Not all swans are white'. Thus a universal hypothesis can be falsified by a basic statement with the logical form of a singular existential statement. This justification of the validity of falsifying arguments has the disadvantage that it only shows how isolated universal statements can be falsified. Perhaps this is why Stegmüller maintains that falsificationism presupposes the testability of isolated hypotheses (Stegmüller 1979, 2: 265). However, it is possible to show that theoretical systems consisting of several general hypotheses can be falsified. In order to show this, I will first deal with the example discussed above of a falsification of a hypothesis with a basic statement. In this example the following prediction is possible:

In the space-time region k there is a swan. (C)
All swans are white. (H)

In the space-time region k there is a white swan. (P)

or

(12) $C, H \vdash P$

This is a very simple prediction using only one hypothesis and one initial condition as premise.

In order to show that falsifications of the second type are valid, I will first show how the isolated hypothesis can be falsified in our example. The following conditional prediction is possible (cf. theorem 1 in the appendix below):

All swans are white. (H)

If there is a swan in the space-time region k, then there is a white swan there. ($C \rightarrow P$)

The negation of the conditional prognosis $\neg(C \rightarrow P)$ is logically equivalent to $C \wedge \neg P$. If the consequence of a valid inference is false, then not all the premises can be true. In this case the conclusion consists of the conditional prognosis ($C \rightarrow P$), the negation of the conclusion is logically equivalent to $C \wedge \neg P$, and there is only one premise, H. Hence the following falsification is valid:

In the space-time region k there is a swan, and in the space-time region k there is not a white swan. ($C \wedge \neg P$)

Not all swans are white. ($\neg H$)

or

(13) $C \wedge \neg P \vdash \neg H$

What is the relation between the falsifying premise $C \wedge \neg P$ and the basic statement B, 'In the space-time region k there is a non-white swan', discussed above? From $C \wedge \neg P$ follows B, but not the other way round. It is at least logically possible that in the space-time region k there is a white and a non-white swan, and hence B can be true and $R \wedge \neg P$ false. Therefore the basic statement B is not logically equivalent to a conjunction of initial conditions and the negation of a prognosis, in spite of Popper's (1959, § 28, 102) view that there is such an equivalence.

Popper's (1959, § 28, 101) formal requirement, that basic statements shall have the logical form of a singular existential statement, is not necessary. He advances it in order to show that falsifications of the second type using basic statements are valid. In his justification for this type of falsification the equivalence between negations of universal statements and existential statements is fundamental. However, the validity of falsifications of the second type can be shown in a much more general way

using metalogical considerations. In order to show the relation between prediction and falsification, it is useful to regard a falsifying basic statement as a conjunction of one or more initial conditions and the negation of an unconditional prognosis, that is as
$C_1 \wedge C_2 \wedge \ldots \wedge C_n \wedge \neg P$,
and not as existential statements of a special kind. The following metalogical equivalence is valid:
(14) $H_1, \ldots, H_r, C_1, \ldots, C_n \vdash P$
if and only if
$C_1 \wedge \ldots \wedge C_n \wedge \neg P \vdash \neg(H_1 \wedge \ldots \wedge H_r)$

It is not necessary that only a single basic statement is used as a premise in falsifications. To make the relation between prediction and falsification as clear as possible, the falsification of 'All swans are white' can be expressed in the following way:

In the space-time region k there is a swan. (R)
In the space-time region k there is no white swan. ($\neg P$)

Not all swans are white. ($\neg H$)

Why are such falsifications valid? The reason is the following metalogical equivalence (theorem 3b in the appendix below):
(15) $H_1, \ldots, H_r, C_1, \ldots, C_n \vdash P$
if and only if
$C_1, \ldots, C_n, \neg P \vdash \neg(H_1 \wedge \ldots \wedge H_r)$
This theorem shows how theoretical systems consisting only of general hypotheses can be falsified. It shows also that in the second type of falsification no "isolationism" has to be assumed, but that whole theoretical systems consisting of several general hypotheses can be falsified.

When the initial conditions describe observable events, there is no reason why they should not belong to the test statements. The methodological requirement that tests should be as severe as possible makes it desirable that the test statements should have as much content as possible. Therefore, not only (unconditional) prognoses, but also observable initial conditions should be regarded as test statements and as possible premises in falsifying arguments.

Initial conditions do not always describe observable events. Some of the initial conditions might be theoretical. Then they must be regarded as parts of the tested system and cannot belong to the test statements. Therefore, the most general logical form of a falsification is the following one:
(16) $C_1, \ldots, C_k, \neg P \vdash \neg(C_{k+1}, \ldots, C_n \wedge H_1 \wedge \ldots \wedge H_r)$ $(0 \leq k \leq n)$
where the initial conditions C_1 to C_k are observable and belong to the test statements and the initial conditions C_{k+1} to C_n are theoretical and belong

to the theoretical system. The following theorem, which is proved as theorem 3c in the appendix, shows that such falsifications are valid:

(17) $H_1, \ldots, H_r, C_1, \ldots, C_n \vdash P$

if and only if

$C_1, \ldots, C_k, \neg P \vdash \neg(C_{k+1} \wedge \ldots \wedge C_n \wedge H_1 \wedge \ldots \wedge H_r)$ $(0 \leq k \leq n)$

In order to test general scientific hypotheses empirically, you first have to derive an observable prognosis from them with the help of initial conditions. If observations show that the prognosis is false, then not all initial conditions and general hypotheses necessary for the derivation of the prognosis can be true. If the derivation of the prognosis is valid, then it follows from the false prognosis that the conjunction of the initial conditions and the general hypotheses cannot be true. This is the first and most simple kind of falsification ($k = 0$ in the theorem above).

When all initial conditions are observable, we get a special kind of falsification. Then a theoretical system consisting only of a conjunction of general hypotheses can be tested and perhaps falsified, as (14) and (15) shows. This type of falsification is a generalisation of the second type of falsification described by Popper (1959, § 28).

These considerations show how closely connected predictions and falsifications are. They also show that a falsification generally hits a whole theoretical system and thus that falsificationism need not presuppose that only isolated hypotheses can be tested and falsified. Deductive logic as an *organon* of criticism is an instrument that can be used in many ways. In some cases it can even be used in order to refute an isolated hypothesis like 'All swans are white'. However, this is only the simplest possible case of a falsification. Usually more complex theoretical systems are tested, systems consisting of many general hypotheses and perhaps also of some theoretical initial conditions, as Duhem (1974, part 2, chap. 6) and Quine (1961, chap. 2) emphasise. However, the systematic or holistic character of science is perfectly compatible with falsificationism. In most cases several hypotheses are necessary in order to derive empirical prognoses. In these cases the falsification hits a theoretical system, not an isolated hypothesis. The *logical* aspects of falsifications are quite clear and unproblematic. Below I will deal with the *methodological* aspects of falsifications. As an introduction to this discussion, I will first deal with some critics of falsificationism.

CHAPTER THREE

KUHN'S HISTORICAL CRITICISM

In order to develop a falsificationist methodology, it is not enough to show that falsifications are logically valid. In order to falsify, you need premises about observed events. The choice of "basic statements" describing such events is problematic if these statements are impregnated by theory, fallible and revisable, as Kuhn maintains.

Usually you do not test isolated hypotheses, but whole theoretical systems. Kuhn asks how falsified theories should be modified, if they should be totally rejected or only slightly modified.

Kuhn's criticism of falsificationism depends on the problem of the empirical "basis" and on the problem of the systematic character of empirical testing. These two methodological problems are basic for his criticism of falsificationism.

A. THEORY AND EXPERIENCE

Kuhn criticises the positivist assumption that there is an infallible and unrevisable empirical basis of science. With many examples from the history of science he shows that the test statements used in science depend on theory and are fallible. Kuhn discusses the following discovery in order to show how observations are determined by accepted theories (or paradigms).

In the history of science it is recorded that Sir William Herschel discovered the planet Uranus in 1781. However, Kuhn points out that Uranus was observed on at least seventeen different occasions between 1690 and 1781, but was held to be a star. After Herschel had discovered that Uranus is a planet, the astronomers saw a planet where they earlier had seen a star, and the accepted test statements changed accordingly. Whereas before 1781 many competent observers reported that there is a star in certain space-time regions, afterwards they reported that there is a planet in corresponding regions (Kuhn 1970b, 115).

How are such changes of test statements possible? In order to explain this, Kuhn refers to psychological experiments showing that the same drawing can be seen in different ways. In *Philosophical Investigations,*

Wittgenstein (1967, part 2, § xi) discusses a drawing that can be seen as a duck or as a rabbit (see fig.1 in chap. 6, B 2, below). Hanson (1958, chap. 1) discusses similar drawings extensively. Inspired by such examples from gestalt psychology, Kuhn maintains that gestalt shifts also occur in science. Scientists learn to see the world in a new way when they discover new things and when they change basic theories. According to Kuhn and Hanson such shifts of perception can be explained with gestalt psychology:

> It is as elementary prototypes for these transformations of the scientist's world that the familiar demonstrations of a switch in visual gestalt prove so suggestive. What were ducks in the scientist's world before the revolution are rabbits afterwards (Kuhn 1970b, 111).

In psychological experiments the same drawing can be seen as a duck or as a rabbit. In a similar way Uranus was earlier seen as a star and later as a planet. Inspired by gestalt psychology Kuhn maintains that Herschel's discovery of Uranus was accompanied by a gestalt switch: before Herschel's discovery the gestalt of a star was seen where later the gestalt of a planet was seen.

According to Kuhn it is not quite correct to say that Uranus was seen *as* a star or *as* a planet. Wittgenstein maintains that normally you do not say that you see a drawing as something, that you see it as a rabbit for example. Instead you say that you see a rabbit in the same way as you say that you see a red circle or knife and fork (Wittgenstein 1967, part 2, § xi). This distinction between 'seeing as' and 'seeing that' was elaborated by Hanson (1958, 19–24), who argued that in science you normally 'see that'. If you say that you see Uranus as a planet or as a star, this suggests that the same observation can be interpreted in different ways. Hanson thinks that this might lead to the positivist assumption that there is a neutral experience that can be interpreted in different ways. However, the perception of a gestalt is immediate and does not contain any conscious interpretation. According to the gestalt psychological model of experience it is therefore more natural to say that the scientists *see that* Uranus is a star or that they *see that* Uranus is a planet. For this reason Kuhn says that after 1781 "there were several fewer stars and one more planet in the world of the professional astronomer" (Kuhn 1970b, 115). Herschel's discovery did not change the interpretation of the world ("seeing as"), but it changed the world itself ("seeing that"), at least the world of the professional astronomer. Scientists do not have any direct access to the world of their research-engagement in itself. "In so far as their only recourse to that world is through what they see and do, we may want to say that after a revolution scientists are responding to a different world" (Kuhn 1970b, 111).

From gestalt psychology Hanson and Kuhn learn that experience itself is theory-dependent. Popper, on the other hand, argues that test *statements* are theory-impregnated and transcend the immediate experience ("transcendence inherent in any description"). Every description is hypothetical and theoretical and goes far beyond what can be verified by immediate experience. For example, the simple statement 'Here is a glass of water' cannot be verified by experience, because it contains universals, like 'glass' and 'water', which cannot be reduced to immediate experience, but "denote physical bodies which exhibit a certain *law-like behaviour*" (Popper 1959, § 25, 95).

In his criticism of the positivist assumption that there is an infallible empirical basis, Popper argues that test statements are theory-impregnated and fallible because they contain universals. Kuhn and Hanson, on the other hand, use gestalt psychology in order to show that experience itself is theory-impregnated and fallible. However, from different points of departure Popper, Hanson, and Kuhn arrive at the same result: the positivist idea that science has an infallible and neutral empirical basis is untenable. All empirical statements, also so called basic statements or test statements, are theory-dependent, fallible, and revisable.

Popper (1979, 45) argues that although a pure "philosophy of immanence" would satisfy the positivist desire to remain in the area of what is immediately given, of what can be immediately experienced, such a philosophy cannot be worked out because it would make all knowledge impossible. Without universals and theories neither experience nor science is possible. Kuhn argues in a similar way with the help of gestalt psychology. The rich experimental literature in this field makes Kuhn suspect that theoretical expectations are prerequisite to perception itself.

> What a man sees depends both upon what he looks at and also upon what his previous visual-conceptual experience has taught him to see. In the absence of such training there can only be, in William James's phrase, "a bloomin' buzzin' confusion" (Kuhn 1970b, 113).

According to Kuhn and James we would face chaos without theoretical presuppositions. However, Popper also thinks that there cannot be any pure experience free from presuppositions, that theoretical presuppositions are necessary not only for science, but also for experience (Popper 1972, 341–61).

However, Popper and Kuhn draw different methodological conclusions from the theory-dependence of experience. Popper thinks that a critical discussion of test statements and theoretical presuppositions is possible, whereas Kuhn denies this. From the theory-dependence of experience Kuhn does not conclude that we are always caught in the same theoretical framework or that theoretical presuppositions (paradigms) or test statements cannot be changed. On the contrary, scientific revolutions and discoveries occur and are accompanied by fundamental changes of observations and test statements. Kuhn compares such changes to gestalt switches. Why do they occur?

Kuhn points to an experiment in gestalt psychology that provides "a wonderfully simple and cogent schema for the process of scientific discovery" (Kuhn 1970b, 64). In that experiment a number of people were asked to identify playing cards on short exposure. "Many of the cards were normal, but some were made anomalous, e.g., a red six of spades and a black four of hearts." Even by short exposures most of the normal cards were correctly identified, but usually the anomalous cards were identified as normal. "Without any awareness of trouble, it was immediately fitted to one of the conceptual categories prepared by prior experience." However, on longer exposures some subjects began to hesitate and to display awareness of anomaly and to the feeling that something was wrong. This confusion caused a kind of crisis of gestalt perception. The subjects did not know how to identify the anomalous cards. However, on still longer exposure most subjects could also identify the anomalous cards, for example a red six of spades (Kuhn 1970b, 62-64).

This experiment is fundamental for Kuhn's conception of science. First I will investigate its consequences for Kuhn's conception of observation. As an example of a scientific discovery Kuhn discusses Herschel's discovery that Uranus is a planet in 1781. Why did Herschel not identify Uranus as a star, as the earlier observers? Why did Herschel experience a gestalt switch causing him to see a planet, where earlier observers had seen a star? When Herschel observed Uranus with a new and much improved telescope in 1781, he saw that Uranus did not look like a point, as stars do in a telescope, but like a little disk. Thus Herschel's improved telescope enabled him to observe an "anomaly" that earlier had not been observed, namely that Uranus looks like a disk. Now a crisis of perception occurred, as in the psychological experiment discussed above. After repeated observations Herschel also saw that Uranus moved daily among the stars. Since stars do not move in this way, this was a new anomaly. Herschel then announced that he had discovered a new comet. Attempting to determine the orbit of Uranus, Lexell discovered that Uranus did not move like a typical comet, but in an almost circular orbit around the sun. He then suggested that Herschel had discovered a new planet. When the astronomers accepted this suggestion, there was a gestalt switch from star to planet. Or, as Kuhn put it, there was one more planet in the world of the professional astronomers (Kuhn 1970b, 115). It is important that this gestalt switch was triggered by "anomalies" that do not fit to a perceived gestalt: the motion and size of Uranus did not fit with the earlier observed gestalt of a star.

When Kuhn discusses the methodological consequences of the theory-dependence of experience, he expresses himself rather vaguely. He does not totally rule out that a neutral observation language might be devised in the future, but only says that modern psychological experimentation is rapidly proliferating phenomena that make it doubtful whether such attempts will succeed. Furthermore Kuhn asks if sensory experience really is so fixed and neutral as the traditional epistemological paradigm suggests. In the absence of a developed alternative paradigm, Kuhn will not relinquish that viewpoint entirely (Kuhn 1970b, 126). Thus Kuhn does not think that he has solved all the problems in the traditional paradigm of fixed and neutral experience. However, Kuhn rules out a position of naive realism, according to which test statements are verified by a process of direct comparison with nature. There is no such higher authority: "The scientist can have no recourse above or beyond what he sees with his eyes and instruments" (Kuhn 1970b, 114). "Under these circumstances we may at least suspect that scientists are right in principle as well as in practice when they treat oxygen and pendulums (and perhaps also atoms and electrons) as the fundamental ingredients of their immediate experience" (Kuhn 1970b, 127–28). The scientists cannot go beyond their gestalt perceptions and find a neutral basis: "The alternative is not some hypothetical 'fixed' vision, but vision through another paradigm" (Kuhn 1970b, 128), for example a paradigm according to which Uranus is a planet, not a star.

Is it possible to compare and criticise different paradigms and statements about observations? Kuhn often uses the metaphor that scientists with different paradigms live in different worlds. For example, when Lavoisier discovered oxygen, he saw oxygen in the same experiments in which Priestly and other adherents of the phlogiston theory saw "dephlogisticated air," that is, air without phlogiston.

> Lavoisier saw nature differently. And in the absence of some recourse to that hypothetical fixed nature that he "saw differently," the principle of economy will urge us to say that after discovering oxygen Lavoisier worked in a different world (Kuhn 1970b, 118).

In a similar way you can say that after the discovery of Uranus Herschel lived in a different world than his predecessors did.

If scientists with different paradigms accept different test statements and even live in different worlds, then different paradigms cannot be compared empirically, but are incommensurable. In this sense Kuhn says that competing paradigms are incommensurable:

> Just because it is a transition between incommensurables, the transition between competing paradigms cannot be made a step at a time, forced by logic and neutral experience. Like the gestalt switch, it must occur all at once ... or not at all (Kuhn 1970b, 150).

Kuhn compares this transition to a conversion and contents himself with describing its psychological and historical aspects. He says that there is no point at which resistance against a new paradigm becomes illogical or unscientific. You can only establish that "if the paradigm is one destined to win its fight" the majority of scientists will accept it (Kuhn 1970b, 159).

In addition, he compares scientific revolutions to political ones: "As in political revolutions, so in paradigm choice—there is no standard higher than the assent of the relevant community" (Kuhn 1970b, 94). Therefore the techniques of persuasive argumentation, the rhetorical elements, have to be examined. There is no "supra-institutional framework for the adjudication of revolutionary difference," and hence "the parties to a revolutionary conflict must finally resort to the techniques of mass persuasion, often including force." Even if scientists do not have to use force in scientific revolution, they have to persuade others. In political and scientific revolutions there is "a choice between incompatible modes of community life." Different paradigms are thus incommensurable forms of scientific life (Kuhn 1970b, 93–94).

How can we explain gestalt switches and changes of experience in science? Do we have to content ourselves with recording such changes historically? Kuhn does not answer these questions. However, it is clear that he thinks that different paradigms are incommensurable and that this raises serious problems for positivism and falsificationism.

What are the consequences of the theory-dependence of experience for falsificationism? If experience depends on theory and is revisable, then test statements are revisable and fallible. Are falsifications at all possible with fallible test statements? Kuhn asks what falsification is if not conclusive disproof and thinks that Popper does not answer this question (Kuhn 1970a, 14–15).

Kuhn interprets falsification as conclusive refutation or disproof. Since test statements are fallible, he thinks that such "apodictic assurance" is impossible. Therefore Kuhn fears that Popper is pursuing a will-o'-the-wisp, the illusion of infallible falsifications, and for the same reason Kuhn finds Popper's use of expressions like 'falsification' and 'refutation' "extremely odd" (Kuhn 1970a, 13).

Basic for Kuhn's discussion is a justificationist interpretation of falsification, according to which falsificiations must be infallible. In this interpretation falsificiations become the negative side of verifications: although a conclusive positive evaluation of theories is impossible, at least the negative evaluation of them should be conclusive in order to make a falsificationist methodology interesting. In Kuhn's interpretation, falsificationism is a negative justificationism, a methodology of conclusive and absolute refutations.

Kuhn correctly states that the theory-dependence of experience makes such a negative justificationism impossible. Since the falsifying premises are fallible and revisable, falsifications themselves must be also fallible and revisable. For this reason Kuhn asks what a falsification is, if not conclusive disproof.

Kuhn knows that Popper also maintains that experience is fallible and theory-dependent. However, he thinks that Popper has not drawn the necessary methodological consequences from this fact, since they would threaten the foundations of a falsificationist methodology:

> Having barred conclusive disproof, he has provided no substitute for it, and the relation he does employ remains that of logical falsification. Though he is not a naive falsificationist, Sir Karl may, I suggest, legitimately be treated as one (Kuhn 1970a, 14).

Kuhn thinks that Popper must assume that test statements are infallible in his methodology and hence that Popper can be treated as a naive falsificationist.

Kuhn maintains that it is not enough to say that theories can be falsified by test statements. A falsification is a logical relation and hence syntactic. However, in actual research logical and syntactic evaluations are not enough, you also have to evaluate what test statements can be maintained on the basis of experience. It is not enough to examine only the relation between statements, you also have to examine the relations between observations and experiments on the one hand and actually asserted test statements on the other. "This is the context in which Sir Karl's term 'falsification' must function, and Sir Karl is entirely silent about how it can do so" (Kuhn 1970a, 15). Put in another way, Kuhn maintains that although Popper has realised that test statements are fallible, he has not told us what kind of fallible test statements actually can be asserted on the basis of experience. Thus Popper has not solved the problem of the "empirical basis." This is the reason why Kuhn maintains that Popper can be treated as a naive falsificationist. Hence Kuhn says that Popper has not provided a "logic" of research, but rather an "ideology" of research, an ideology of conclusive refutations with infallible test statements, an ideology of negative justificationism. The favourable response such an ideology has found among scientists can be explained by the fact that scientists accept paradigms and test statements uncritically and hence think that test statements and falsificiations are infallible.

B. Normal and Extraordinary Science

Kuhn distinguishes between normal and extraordinary science. Using an example from the history of science, I will fist deal with normal science. Kuhn (1957, 73–77) asks why the Copernican revolution occurred. Was it caused by a falsification of the earlier Ptolemaic theory? In order to discuss this question Kuhn describes what he calls the logical structure of a scientific revolution in the following way: "A single observation incompatible with his theory demonstrates that he [the scientist] has been employing the wrong theory all along. His conceptual scheme must then be abandoned and replaced" (Kuhn 1957, 75).

Kuhn does not think that these logical considerations are sufficient in order to explain a scientific revolution:

> Historically the process of revolution is never, and could not possibly be, so simple as the logical outline indicates. As we have already begun to discover, observation is never *absolutely* incompatible with a conceptual scheme (Kuhn 1957, 76).

Why are theory and observations never absolutely incompatible? The reason is not that test statements are fallible, but that falsified theories can be modified. Therefore scientists do not always have to abandon a falsified theory completely or replace it with a new one. In his discussion of the Copernican revolution Kuhn assumes that the Ptolemaic theory also in its most developed form "did not precisely reconcile theory and observation" (Kuhn 1957, 74) and hence was falsified from a logical point of view. However, since the Ptolemaic theory was a complex theoretical system with many component hypotheses about deferents, eccentrics, epicycles, and equants, it was not necessary to abandon the theory completely and replace it with a new one. You could also try to modify the old theory and in this way reconcile theory and observation, for example by changing some peripheral hypotheses of the theory or by adding a new auxiliary hypothesis. Even if the Ptolemaic theory was falsified, it was not necessary to abandon it completely. Kuhn thinks that you disregard an important problem if you concentrate on logical relations: Why do scientists sometimes try to solve problems by modifications of available theories and sometimes by introduction of new theories? Why did Copernicus introduce a new theory instead of modifying the Ptolemaic theory?

> What is it that transforms an apparently temporary discrepancy into an inescapable conflict? ... Why do scientists hold to theories despite discrepancies, and, having held to them, why do they give them up? These are problems in the anatomy of scientific belief (Kuhn 1957, 76).

Kuhn's questions depend on the fact that different research strategies are possible when a theory is falsified. You can use a revolutionary strategy and try to introduce quite a new theory, as Copernicus did, or you can use a more cautious strategy and try to solve problems by modifying the available theory.

Kuhn assumes that a falsificationist methodology requires that falsified theories are always completely abandoned, that a revolutionary research strategy always has to be used. Science does not develop in this way, as Kuhn shows with many examples from the history of science. Instead, often a more cautious strategy is used, and available theories modified more or less extensively (Kuhn 1970b, chaps. 3–4).

In order to explain the two research strategies, the normal and the extraordinary one, Kuhn calls attention to the beliefs of the scientists. If a scientist believes in an available theory, he will use the cautious research strategy. Only when he does not believe in it, will he use the revolutionary research strategy. According to Kuhn, the Copernican revolution occurred because unsolved problems had shattered Copernicus' belief in the Ptolemaic theory:

> An honest appraisal of contemporary astronomy, says Copernicus, shows that the earth-centered approach to the problem of the planets is hopeless. The traditional techniques of Ptolemaic astronomy have not and will not solve that problem; ... there must, he concludes, be a fundamental error in the basic concepts of traditional planetary astronomy (Kuhn 1957, 139).

Thus Kuhn thinks that Copernicus introduced a new theory because he no longer believed that the motions of the planets could be satisfactorily explained by the Ptolemaic theory.

In *The Structure of Scientific Revolutions* Kuhn develops his ideas on the different research strategies further. There he calls the cautious strategy of small theoretical modifications "normal science" and the revolutionary strategy "extraordinary science" and develops a model for the historical development of science. If there is a fundamental tradition of research and a basic paradigm, then the strategy of "normal science" is used. Normal science continues as long as the paradigm functions well and scientists solve emerging problems or, as Kuhn calls them, puzzles. When scientists fail to solve puzzles with a paradigm and anomalies pile up, then they lose faith in the paradigm and a crisis occurs. In the relatively rare crises scientists search for new theories, and many theories or paradigms may compete. However, this phase of extraordinary science ends when a successful and promising paradigm is found. Then the phase of normal science begins again and will continue until a new crisis occurs. This is Kuhn's model for the development of science. It maintains that science is normally governed by fixed traditions and unquestioned paradigms. Only in extraordinary crises are alternative paradigm considered and scientific revolutions occur (Kuhn 1970b, chap. 9).

With his model Kuhn criticises Popper's falsificationist view of science. He thinks that Popper only deals with the rare moments of extraordinary science, when new theories emerge, but not with the more prosaic moments of normal science, when paradigmatic theories are used as instruments in order to solve puzzles. According to Kuhn, Popper's methodology depends on a misleading generalisation of extraordinary and rare moments of science.

Contrary to Popper, Kuhn argues that in normal science scientists do not test their theories critically, but believe in them and use them as instruments in order to solve problems. In normal science scientists regard problems as 'puzzles' to be solved within the generally accepted paradigm.

Not all problems can be regarded as puzzles. When observations contradict a theory, then these observations cannot be explained by the theory without further ado. In order to find an explanation, you must at least modify the theory. Kuhn calls problems of this kind anomalies. They depend on observations that do not fit into the available theory. Problems of this kind can always emerge. They are not sufficient for a scientific revolution and can often be assimilated by small modifications of the available theory (Kuhn 1970b, chap. 4)

In normal science problems do not challenge the theory, but the acumen of the scientist. Kuhn compares such problems to jigsaw puzzles and crossword puzzles, which also have the function of testing ingenuity and skill and which have the assured existence of a solution. Therefore the term 'puzzle' is especially appropriate for the problems of normal science (Kuhn 1970b, 36–37).

In normal science scientists try to solve standard problems with standard methods. If they do not succeed in finding solutions in this way, if the puzzles remain unsolved and the anomalies unassimilated, then these failures do not hit the paradigm, but the scientist, who was not ingenious enough to find a solution with the accepted paradigm. The inability to find a solution discredits the scientist, not the paradigm. If such a scientist should blame the paradigm for his failure, he would inevitably be seen by his colleagues as "the carpenter who blames his tools" (Kuhn 1970b, 79). In normal science it thus discredits a scientist to maintain that paradigms have serious defects.

Kuhn shares this attitude with the conventionalist philosophers of science. As Popper writes, these philosophers "cast doubt on the acumen of the theoretician" who is unable to solve problems with an accepted theory. Popper calls this a conventionalist objection against falsificationism and does not find it important (Popper 1959, § 19, 81). It is a mere expression of belief in the paradigm.

If Kuhn's description of normal science is correct, then falsificationism can be criticised in the following way. In normal science scientists are not critical, but are governed by traditions and paradigms. They do not test paradigms and do not try to refute them, but use them as instruments in order to solve problems. In spite of "anomalies" the paradigm is retained and at the most modified. Paradigms are not tested critically, but are necessary presuppositions for normal science and its "puzzle-solving." In normal science scientists do not try to discover new phenomena or new theories (Kuhn 1970b, 35).

According to Kuhn discrepancies between predictions and observations do not lead to rejection of theories in normal science. Therefore he (Kuhn 1970b, 146) maintains that "anomalous experiences may not be identified with falsifying ones. Indeed, I doubt that the latter exist." When Kuhn doubts that falsifying experiences exist, he does not doubt that test statements can contradict theories logically. This would be an absurd opinion. I suppose that Kuhn wants to say that there cannot be any test statements forcing us to reject a theory completely (cf. Kuhn 1957, 76–77).

Kuhn thinks that there are always unsolved puzzles and perhaps also unassimilated anomalies. The data-theory fit is always imperfect: "If any and every failure to fit were ground for theory rejection, all theories ought to be rejected at all times" (Kuhn 1970b, 146). However, it is unclear what Kuhn means with 'puzzle' and 'failure to fit'. One type of scientific problem arises when there are unexplained observations that do not contradict a theory. Another type of problem arises when the observations are not only unexplained, but also contradict a theory. It might seem as if Kuhn does not want to say only that there are always unexplained observations, but also that there are always contradictions between observation and theory: "there is no such thing as research without counterinstances" (Kuhn 1970b, 79). However, Kuhn's "counterinstances" probably do not always contradict a theory: "Excepting those that are exclusively instrumental, every problem that normal science sees as a puzzle can be seen, from another viewpoint, as a counterinstance and thus as a source of crisis" (Kuhn 1970b, 79). All problems and puzzles need not necessarily contradict a particular theory. This vagueness would not exist if Kuhn conceived counterinstances in a logical way. In logic either there is a contradiction or not. Since Kuhn conceives counterinstances in a heuristic way, there are different possibilities: scientists can try to solve problems in a normal or extraordinary way. Kuhn's discussion of puzzles, anomalies, and counterinstances is unclear because he does not make any logical distinctions. Kuhn probably thinks that such distinctions are unimportant compared to psychological considerations. If an observation is psychologically unexpected, then Kuhn calls it an anomaly. If a scientist tries to assimilate an anomaly with the accepted paradigm, then the anomaly is regarded as a puzzle. If the anomaly cannot be assimilated in this way, the scientist loses faith in the paradigm and the anomaly causes a crisis. Not all anomalies lead to crises; many can be assimilated in the normal way by more or less extensive modifications of the accepted paradigm. If a new paradigm is suggested in a crisis and if the new paradigm seems to solve problems better than the old one, then the anomaly can become a counterinstance causing a scientific revolution and the acceptance of a new paradigm (Kuhn 1970b, chaps. 9 and 12).

Since anomalies can be regarded as puzzles or as counterinstances, Kuhn confronts falsificationism with the following dilemma: "either no scientific theory ever confronts a counterinstance, or all such theories confront counterinstances at all times" (Kuhn 1970b, 80). If anomalies are always regarded as puzzles, theories never confront counterinstances; if anomalies are always regarded as counterinstances, theories always confront counterinstances. Kuhn does not think that a falsificationist methodology can explain why problems can be regarded so differently. Kuhn's own explanation is psychological: when the belief in the paradigm is unshaken in normal science, problems are regarded as puzzles; when the belief in the paradigm is shaken in a crisis and there is an alternative paradigm that seems better, problems are regarded as counterinstances. A theory is abandoned first when two conditions are met: (1) there must be a crisis undermining the belief in the ruling paradigm and making the search for an alternative paradigm possible, (2) as a result of the crisis there must be an alternative paradigm that seems to be more promising. Without an apparently better paradigm an old paradigm is never abandoned and problems are never seen as counterinstances. Although severe and prolonged anomalies may trigger a crisis and may undermine the faith in a paradigm, they are not regarded as counterinstances without an alternative paradigm:

> once it [a paradigm] has achieved the status of paradigm, a scientific theory is declared invalid only if an alternate candidate is available to take its place. ... The decision to reject one paradigm is always simultaneously the decision to accept another (Kuhn 1970b, 77).

It is an instrumentalist idea that a scientist rejects a theory first when another and apparently better theory is available.

Kuhn (1970b, 77) concludes that falsificationism depends on the "methodological stereotype" of falsification by direct comparison with nature. He does not think that falsificationism contributes much to the understanding of science and its history. Impressed by this criticism many philosophers of science thought that falsificationism is historically and methodologically untenable.

CHAPTER FOUR

LAKATOS'S ANSWER TO HISTORICAL CRITICISM

Imre Lakatos tries to answer Kuhn's historical criticism of falsificationism. He regards Kuhn's own position as relativist and summarises it in the following way:

> For Kuhn scientific change—from one 'paradigm' to another—is a mystical conversion which is not and cannot be governed by rules of reason and which falls totally within the realm of the *(social) psychology of discovery*. Scientific change is a kind of religious change (Lakatos 1970, 93).

Lakatos (1970, 91) asks whether Kuhn's view is correct, whether science is "reason or religion." Less provocatively formulated, Lakatos asks whether a rational philosophy of science is possible, or whether we have to content ourselves with psychological explanations of the development of science. When Lakatos asks whether science is rational, he does not ask in the first instance whether the history of science can be "rationally reconstructed," but rather whether a rational normative methodology can be defended. His discussions of falsificationism and of the methodology of scientific research programmes are methodological discussions. Lakatos is much impressed by Kuhn's criticism of falsificationism and thinks that Kuhn has shown that important parts of falsificationism can be criticised with arguments from the history of science. However, Lakatos tries to show that in Popper's methodology two different positions are conflated, a naive and a sophisticated falsificationism. For the naive falsificationism Kuhn's criticism is valid, but not for the sophisticated one, which Lakatos develops to his own methodology of scientific research programmes. However, Lakatos first deals with a kind of falsificationism that is dogmatic and even more naive than naive falsificationism.

A. DOGMATIC FALSIFICATIONISM

Lakatos (1970, 95) calls dogmatic falsificationism "a most important brand of falsificationism," which, however, Popper has never published anything about. Nevertheless, Lakatos thinks that Popper began as a dogmatic falsificationist in the twenties, but published nothing until he had invented methodological falsificationism (Lakatos 1970, 180). Lakatos does not tell us how he knows that Popper was ever a dogmatic falsificationist. Since the alleged dogmatic falsificationist published nothing, this is a mystery. In the meantime, Popper's preparatory work to *The Logic of Scientific Discovery* has been published (Popper 1979). There you cannot find any dogmatic falsificationism. Therefore it is advisable to regard dogmatic falsificationism as Lakatos's own invention and as having nothing to do with Popper. Dogmatic falsificationism is, however, interesting as a usual justificationist misunderstanding of Popper's falsificationism.

Characteristic for dogmatic falsificationism is the assumption that test statements and falsifications are infallible, that "there is an absolutely firm empirical basis of facts which can be used to disprove theories" (Lakatos 1970, 96). Dogmatic falsificationism is not an uncompromising fallibilist position, since test statements and falsifications are assumed to be infallible. It is rather a justificationist falsificationism: although it does not assume that verifications and positive justifications of universal theories are possible, it nevertheless assumes that falsifications are infallible and conclusive refutations or disproofs. Although Lakatos does not mention it, Kuhn (1970a, 14–15) interprets Popper as a dogmatic falsificationist.

Lakatos does not think that infallible test statements are enough for infallible falsifications. Even if test statements were infallible, "their disproving power would still be miserably restricted: *exactly the most admired scientific theories simply fail to forbid any observable state of affairs* (Lakatos 1970, 100). Lakatos thinks that it is always possible to introduce auxiliary hypotheses in order to immunise a theory against criticism and make it agree with experience. To show this, Lakatos discusses the following imaginary case. He assumes that the orbit of a planet has been calculated using Newton's mechanics and law of gravitation, but that the planet deviates from the calculated path. From a logical point of view this is a falsification. However, Lakatos does not think that a "Newtonian physicist" would discard Newton's theory for this reason, but that he would introduce auxiliary hypotheses in order to save it. He could maintain, for example, that the movements of the planet are disturbed by another, until now unknown planet. Even if the assumed perturbing planet cannot be observed, the "Newtonian physicist" would not regard his theory as false. Instead he would introduce another auxiliary hypothesis, for example that the assumed planet is so small that it cannot be observed or that "a cloud of cosmic dust hides the planet from us." So the "Newtonian physicist" could always continue to introduce new auxiliary hypotheses in order to immunise his theory from criticism, or the whole story would be buried in the dusty volumes of periodicals and never mentioned again (Lakatos 1970, 101).

Lakatos's example shows how falsifications can be immunised by the introduction of new auxiliary hypotheses. Conventionalist philosophers of science argued that falsifications can always be immunised in this way. This is an interesting objection to falsificationism, which Popper (1959, § 19) calls a conventionalist objection. Lakatos thinks that this objection has the consequence that all theories are not only unverifiable, but also unfalsifiable, and hence that dogmatic falsificationism is untenable: "*Scientific theories are not only equally unprovable, and equally improbable, but they are also equally undisprovable*" (Lakatos 1970, 103).

The conventionalists also maintained that test statements are theory-dependent and fallible. Lakatos (1970, 95–100) regards this as a further objection to dogmatic falsificationism and the idea of disproof of theories. Both objections to falsificationism, the possibility to immunise theories against criticism with auxiliary hypotheses and the fallibility of test statements, are dealt with by Popper in *The Logic of Scientific Discovery* (Popper 1959, § 19, chap. 5). Therefore, Lakatos (1970, 181, fn. 2) calls this book "the strongest ever criticism of dogmatic falsificationism."

B. NAIVE METHODOLOGICAL FALSIFICATIONISM

The fallibility of test statements and possibility of immunising theories against criticism are important problems in a falsificationist methodology. Which fallible test statements can be used in order to falsify theories? How do we restrain the introduction of auxiliary hypotheses used in order to immunise criticism? Popper's answers to these questions in *The Logic of Scientific Discovery* are fundamental for what Lakatos calls "methodological falsificationism," which he, curiously enough, thinks is a brand of conventionalism (Lakatos 1970, 104).

Fundamental for methodological falsificationism is a new solution to the problem of test statements. They can no longer be proved through experience, as in dogmatic falsificationism, but are made unrefutable through a conventionalist decision: "Our *revolutionary conventionalist* (or 'methodological falsificationist') makes unfalsifiable by *fiat* some (spatio-temporally) singular statements which are distinguishable by the fact that there exists at the time a 'relevant technique' such that 'anyone who has learned it' will be able to *decide* that the statement is 'acceptable'" (Lakatos 1970, 106). Thus Lakatos interprets Popper as saying that test statements are made unfalsifiable by conventionalist decisions. The decision to accept test statements as true is insufficient. You also have to decide which statements are test statements: "Indeed, the very selection of all such statements [as test statements] is a matter of decision, which is not based on exclusively psychological consideration" (Lakatos 1970, 106). In order to lay down an "empirical basis" of test statements, two decisions are necessary: (1) the decision to regard special statements as test statements, (2) the decision to regard some test statements as true. The first decision is at the same time a decision to regard special events as observable, since test statements should describe observable events (Popper 1959, § 28).

The conventional decisions about test statements are insufficient. In order to make probabilistic theories falsifiable, a third kind of decision is necessary making "statistically interpreted evidence 'inconsistent' with the probabilistic theory" (Lakatos 1970, 109).

In order to falsify ordinary nonprobabilistic theories an additional decision is often necessary. Lakatos (1970, 101–2, 110–11) first thought that some theories cannot explain anything "observable" without a *ceteris paribus* clause. The reason is that "some scientific theories forbid an event ... only on the condition that no other factor (possibly hidden in some distant and unspecified spatio-temporal corner of the universe) has any influence on it" (Lakatos 1970, 101 [without Lakatos's italics)]. Such a condition is expressed in a *ceteris paribus* clause. In Lakatos's example discussed earlier the Newtonian theory was tested and immunised against criticism by the introduction of auxiliary hypotheses, for example by the hypothesis that an undiscovered planet perturbs the orbit of an observed planet. Lakatos wants to rule out such immunising auxiliary hypotheses by *ceteris paribus* clauses. Their function is to rule out immunisation of theories against criticism and to make falsifications infallible and absolute. The quest for absolutely infallible falsifications is as strong in methodological falsificationism as in dogmatic falsificationism: the "conventional decisions" about test statements and *ceteris paribus* clauses have the function of making falsifications infallible.

Lakatos regards the decision to accept a *ceteris paribus* clause as a very risky one. With a *ceteris paribus* clause mere "anomalies" become "crucial evidence" against a theory, that is they become absolute falsifications. "Since, for our savage [methodological] falsificationist, falsificiations are methodologically conclusive, the fateful decision amounts to the methodological elimination of ... [a] theory, making further work on it irrational" (Lakatos 1970, 111). Without a *ceteris paribus* clause a falsification is a mere anomaly and does not lead to the abandonment or elimination of the tested theory, which can be protected by appropriate auxiliary hypotheses. By means of *ceteris paribus* clauses Lakatos hopes to elucidate Kuhn's distinction between anomalies and counterexamples.

Like Kuhn, Lakatos thinks that falsifications must be conclusive and make further work on the falsified theory irrational.

Lakatos (1970, 110) first maintains that a *ceteris paribus* clause is necessary in order to derive predictions from a theory. This is a logical blunder, as Lakatos (1970, 186) later realises. However, this insight does not reduce the number of "conventional decisions" in methodological falsificationism, at least not in its "naive" variant, because Lakatos replaces decisions about *ceteris paribus* clauses by decisions about "uncontested background knowledge." In this way a conclusive falsification of a theory becomes possible by a conventional decision about background knowledge (Lakatos 1970, 186).

Lakatos (1970, 112) introduces an additional decision in order to make metaphysical theories falsifiable. A metaphysical theory is a theory that cannot be contradicted by test statements. However, in some cases it is contradicted by a scientific theory. In these cases Lakatos decides by a fifth type of conventional decision to "eliminate" the metaphysical theory and to regard it as falsified. It is unclear why this type of decisions is necessary. If there is a contradiction between a metaphysical and a scientific theory and if we assume that the scientific theory is true, then it follows logically and without any conventional decision that the metaphysical theory is false. Perhaps this type of decision is necessary in order to make the falsification methodologically conclusive, in order to make further work on the refuted theory irrational. (The decision about *ceteris paribus* clauses had this function.) It is, however, problematic whether conventional decisions can make it irrational to continue to work on a theory.

Lakatos says that decisions play a crucial role in naive methodological falsificationism. He (1970, 112) finds this unsatisfactory since "decisions ... may lead us disastrously astray." Therefore Lakatos (1970, 114) asks: "Is not the firm strategy of the brand of methodological falsificationism hitherto discussed *too firm*? Are not the decisions it advocated bound to be *too arbitrary*?"

The decisive argument against naive methodological falsificationism Lakatos (1970, 114) finds in the history of science: "If we look at history of science, if we try to see how some of the most celebrated falsificiations happened, we have to come to the conclusion that either some of them are plainly irrational, or that they rest on rationality principles radically different from the one we just discussed." As examples Lakatos (1970, 115) states that "eighty-five years elapsed between the acceptance of [the movement of] the perihelion of Mercury as an anomaly and its acceptance as a falsification of Newton's theory" and that "Galileo and his disciples accepted Copernican heliocentric celestial mechanics in spite of the abundant evidence against the rotation of the Earth." These examples have the same function as some of Kuhn's historical cases. They are intended to show that theories are not eliminated although they are falsified by test statements. Such cases make Lakatos think that new rules of elimination or falsification have to be introduced. Only a naive falsificationist eliminates a theory refuted by experience without further ado. A sophisticated falsificationist proceeds in a different way.

C. SOPHISTICATED METHODOLOGICAL FALSIFICATIONISM

In naive methodological falsificationism Popper's criterion of demarcation is still valid, and a theory is regarded as scientific if it is falsifiable. In sophisticated methodological falsificationism Lakatos suggests a new criterion of demarcation. According to this criterion a theory *T´* is scientific if and only if its predecessor *T* has the following characteristics: (1) *T´* has excess empirical content over *T*, and (2) some of the excess content of *T´* is corroborated (Lakatos 1970, 116).

According to Lakatos a naive methodological falsificationist eliminates a falsified theory. A sophisticated falsificationist, on the other hand, only eliminates a theory *T* if there is another theory *T´* such that the conditions (1) and (2) above and an additional condition (3) are satisfied: "*T´* explains the previous success of *T*, that is, all the unrefuted content of *T* is contained (within the limits of observational error) in the content of *T´*" (Lakatos 1970, 116).

Is the third condition really necessary? Must not a theory with excess empirical content in relation to its predecessor necessarily explain all the previous successes of its predecessor? This is not the case because Lakatos (1968a, 390–495; 1970, 114) defines 'excess empirical content' in a quite special way: *T´* has excess empirical content in relation to *T* if and only if *T´* has potential falsifiers that are not potential falsifiers of *T*. Lakatos does not require that all the potential falsifiers of *T* are also potential falsifiers of *T´*. This has the paradoxical consequence that *T´* might have excess empirical content in relation to *T* at the same time as *T* has excess empirical content in relation to *T´*, as Lakatos (1968a, 379) admits. According to ordinary usage, there is an excess of empirical content if there is *more* empirical content, not if there is *different* empirical content. Lakatos's peculiar use of 'excess empirical content' has its reasons. In *The Logic of Scientific Discovery* there is no general solution to the problem of comparing the empirical content or degree of testability of different theories. Popper only deals with special cases, for example with the case when the empirical content of one theory is a subset of the empirical content of another. In other cases the empirical contents of the theories cannot be compared, but are incommensurable

(Popper 1959, § 32). By his odd definition of 'excess empirical content' Lakatos avoids this problem.

In sophisticated methodological falsificationism a theory is "falsified" when the three conditions above are satisfied. This has the paradoxical consequence that a true theory might be "falsified." Assume that the true theory *T´* is logically stronger than its predecessor *T*, that is, that *T* logically follows from *T´*. Then *T* must be true. Assume further that some true (or "corroborated") new predictions follow from *T´*, but not from *T*. According to Lakatos's new rules of elimination, the true theory *T* is "falsified," since (1) *T´* has excess empirical content in relation to *T*, (2) a part of this content is corroborated, and (3) the logical stronger theory *T´* can explain all the previous successes of *T*.

It would have been better if Lakatos had only said that in sophisticated methodological falsificationism theories are eliminated. It is totally redundant to call such theories falsified. Actually, normal logical falsifications do not play any role whatsoever in sophisticated falsificationism, but are rejected for historical reasons. After having read Kuhn, Lakatos thinks that in the historical development of science many (in the logical sense) falsified theories are not rejected. For this reason Lakatos introduces new "rules of elimination" in sophisticated falsificationism and totally gives up Popper's fundamental idea of logical falsification. This explains why Lakatos (1970, 122, fn. 4) thinks that in future it might be better to abandon terms like refutation, falsification, and counterexample altogether. Sophisticated falsificationism is falsificationism without falsifications. Perhaps this is the reason why Lakatos calls it sophisticated.

The best way to understand Lakatos's rules of elimination is to see them in relation to Kuhn's ideas. Kuhn maintained that a theory is abandoned (or eliminated) first when there is another theory that appears to be better. However, Kuhn thinks that it is impossible to give objective criteria for the quality of theories. Hence he sees the transition to a new theory as a conversion (see chap. 3, B, above). Just this point is crucial for Lakatos, who thinks that it is possible to give objective criteria for the quality of theories and even of research programmes. Lakatos recommends that if there is an objectively better theory, an inferior theory should be eliminated.

Lakatos's rules of elimination are similar to Popper's three requirements for the growth of knowledge. According to Popper's (1965, 241) second requirement, a new theory should be independently testable: "Apart from explaining all the *explicanda* which the new theory was designed to explain, it must have new and testable consequences (preferably consequences of a *new kind*); it must lead to the prediction of phenomena which have not so far been observed." Popper and Lakatos speak about new phenomena in different ways. For Popper a phenomenon is new if it has not earlier been observed, while for Lakatos it is new if it cannot be predicted by an earlier theory. According to Popper's (1965, 242) third requirement for the growth of knowledge "we require that the [new] theory should pass some new, and severe, tests." The conjunction of Popper's second and third requirement is equivalent to Lakatos's rule of elimination, although Lakatos's rule has quite another methodological function than Popper's requirements. Popper uses them for conjectures about the growth of knowledge, while Lakatos uses them in order to eliminate theories conclusively.

In sophisticated methodological falsificationism Lakatos deals with the problem that falsified theories are not always totally rejected and replaced by quite new ones, but often more or less extensively modified. Earlier, in naive methodological falsificationism, Lakatos did not allow the modification of falsified theories, but required their total elimination. This research strategy disagrees with the historical development of science and does not always contribute to the growth of knowledge. Therefore Lakatos allows scientists to continue to work on and to modify falsified theories in sophisticated methodological falsificationism. Lakatos wants to make what Kuhn called normal science respectable and rational.

Kuhn criticised falsificationism by pointing at the theory-dependence of experience and the incommensurability of competing theories, but Lakatos does not discuss these points so extensively. In a sense, test statements are fallible in sophisticated methodological falsificationism, whereas they were made unfalsifiable by conventional decisions in the naive variant. In sophisticated methodological falsificationism Lakatos (1970, 127) introduces an *appeal procedure* against the decision to accept test statements. How is such criticism of test statements possible? Lakatos (1970, 127–28) first examines Popper's idea that test statements can be tested by their deductive consequences (Popper 1959, § 29), but rejects it because he thinks that normally test statements are criticised indirectly by their "interpretative theory" rather than by their deductive consequences: "when the theoretician appeals against the verdict of the experimentalist, the appeal court does not normally cross-question the basic statement directly but rather questions the *interpretative theory* in the light of which its truth-value had been established" (Lakatos 1970, 128). How are interpretative theories cross-questioned in Lakatos's appeal court? They are questioned by attempts to replace the interpretative theory, or the tested theory, or possibly both, and then the scientists "opt to that new set-up which provides the biggest increase in corroborated content" (Lakatos 1970, 130).

Lakatos's solution to the problem of how to criticise test statements is circular. Test statements are criticised with the help of interpretative theories, but interpretative theories are ultimately judged by their empirical content, that is by test statements. First the empirical content of a theory is determined with the help of an interpretative theory, but at the same time interpretative theories are judged by their empirical content. If Lakatos had been consistent, the idea that the truth values of test statements are determined with the help of interpretative theories would have led to what Kuhn called the incommensurability of competing theories, to the consequence that with different interpretative theories different test statements and empirical contents are accepted. Lakatos does not have any independent and neutral appeal court in order to cross-examine test statements. Referring to Popper (1966, 218), Lakatos (1970, 131) says that experience remains the "impartial arbiter" of scientific controversies. In Lakatos's sophisticated methodological falsificationism experience is no longer impartial, but heavily influenced by interpretative theories. If Lakatos had taken his idea of interpretative theories more seriously, different theories (and different research programmes) would usually have different interpretative theories and hence different empirical contents, that is, competing theories would usually be incommensurable. This would have ruined Lakatos's attempt to save reason in science.

In sophisticated methodological falsificationism decisions about *ceteris paribus* clauses are no longer necessary because Lakatos no longer thinks that such clauses are needed in order to explain events. Instead he now thinks that they are explained with universal hypotheses, auxiliary hypotheses, and initial conditions. If the predicted event does not occur, then the theoretical system consisting of these universal hypotheses, auxiliary hypotheses, and initial conditions is falsified. In his discussion of naive methodological falsificationism Lakatos suggested that a (conventional) decision is necessary in order to know what part of the theoretical system is falsified. In sophisticated methodological falsificationism such a decision is no longer necessary: "We may try to replace *any* part and only when we have hit on an explanation of the anomaly with the help of some content-increasing change ..., and nature corroborates it, do we move on to eliminate the 'refuted' complex" (Lakatos 1970, 125). Here Lakatos deals with what was called the first type of falsification (see chap. 2, C, above) , that is, with falsification of a theoretical system with the help of the negation of a prognosis. Such falsifications have the logical form (cf. (11) in chap. 2, C 1, above) :

(1) $\neg P \vdash \neg(H_1 \wedge \ldots \wedge H_r \wedge \ldots \wedge C_1 \wedge \ldots \wedge C_n)$

According to De Morgan's law (1) is logically equivalent to:

(2) $\neg P \vdash \neg H_1 \vee \ldots \vee \neg H_r \vee \ldots \vee \neg C_1 \vee \ldots \vee \neg C_n$

This type of falsification hits a whole theoretical system. The conjunction of all the parts of the system cannot be true. The falsification does not hit any specific part of the system: it does not tell us which parts of the system are false. In sophisticated methodological falsificationism it is not necessary to decide which parts are false: "we do not have to decide which ingredients of the theory we regard as problematic and which ones as unproblematic." Instead we try to replace some ingredients and then test the modified system. "If we succeed in replacing some ingredient in a 'progressive' way (that is, the replacement [the modified theoretical system] has more corroborated empirical content than the original [theoretical system]), we call it [the original] 'falsified'" (Lakatos 1970, 125).

However, in naive methodological falsificationism decisions are necessary in order to know which parts of a falsified system are false: "The naive falsificationist insists that if we have an inconsistent set of scientific statements, we first must select from among them (1) a theory under test (to serve as a *nut*); then we must select (2) an accepted basic statement (to serve as a *hammer*) and the rest will be uncontested background knowledge (to provide an *anvil*)" (Lakatos 1970, 186). By such decisions or choices isolated hypotheses can be tested and falsified (isolationism). Such falsificiations depend on decisions to accept some initial conditions as true and some general hypotheses as belonging to unproblematic background knowledge.

Why are such decisions necessary? The "Quine-Duhem thesis" denies in its weak interpretation "the possibility of *disproof* of any *separate* part of a theoretical system" and in its strong interpretation it "excludes any *rational* selection rule among the alternatives" (Lakatos 1970, 184). Thus the "Quine-Duhem thesis" in its weak interpretation presupposes that whole theoretical systems (consisting of initial conditions and universal hypotheses) are tested empirically. In its strong interpretation it denies that you can decide in a rational way which parts of a refuted system are false. The last problem can be called the Quine-Duhem problem. In naive methodological falsificationism this problem is solved by deciding conventionally which parts of the system belong to the unproblematic background knowledge. In sophisticated methodological falsificationism Lakatos hopes to solve the problem by giving rational rules saying that falsified systems should be modified in a content-increasing way. If a falsified theoretical system is modified in such a way that the new system has "excess empirical content" and some of this content is corroborated, then the modified system is scientific (and not *ad hoc*). If the modified system also explains the successes of the earlier system, then the earlier system should be eliminated as "falsified." Thus you judge whole theoretical systems in sophisticated methodological falsificationism, not isolated hypotheses, and you do not have to decide which parts of a falsified system are false.

In order to discuss the Duhem-Quine problem, Lakatos (1970, 185) deals with Popper's example of a scientific explanation in the following way:

> Let h_1 be: 'whenever a thread is loaded with a weight exceeding that which characterises the tensile strength of the thread, then it will break'; let h_2 be: 'the weight characteristic for this thread is 1 *lb*'; let h_3 be: 'the weight put on this thread was 2 *lbs*'. Let, finally, h_0 be: 'an iron weight of 2 *lbs* was put on the thread located in the space-time position P and it did not break'.

This falsification has the following logical form:

(3) $h_0 \vdash \neg(h_1 \wedge h_2 \wedge h_3)$

Lakatos's statement h_3 is not a universal hypothesis, but a singular initial condition. It is problematic whether the second hypothesis (h_2) is a universal hypothesis or an initial condition. As far as it is saying something about "this thread," it is singular; as far as it is saying something about the tensile strength of this thread in all space-time-regions, it is universal. In the following I will regard h_2 as a hypothesis. In Popper's original example this difficulty does not emerge since Popper formulates a general hypothesis about the relation between tensile strength and the structure of threads.

Lakatos's example can be reconstructed as:

> Whenever a thread is loaded with a weight exceeding its tensile strength, then it will break. (Hypotheses$_1$: H_1)
> The tensile strength of this thread is 1 *lb*. (Hypothesis$_2$: H_2)
> The weight put on this thread in the space-time position k was 2 *lbs*. (Initial condition$_1$: C_1)
> ___
> The thread in the space-time position k breaks. (Prognosis: P)

Or:

(4) $H_1, H_2, C_1 \vdash P$

The falsification intended by Lakatos has the following logical structure (cf. (11) in chap. 2, C 1, above):

(5) $\neg P \vdash \neg(H_1 \wedge H_2 \wedge C_1)$

When a theory is falsified, Lakatos (1970, 185) thinks that we can avoid the falsification by changing any of the sentences in the falsifying inference. So we can change H_1 into: "If a thread is pulled by a force exceeding its tensile strength, then it will break" (H'_1). Moreover, a new initial condition can be introduced: "There was a hidden magnet (or hitherto unknown force) located in the laboratory ceiling." Although Lakatos does not mention it, further changes of the theoretical system are necessary in order to avoid a falsification. It is, for example, necessary to specify the strength of the hidden magnet. In this way Lakatos exemplifies the conventionalist thesis that through the introduction of appropriate auxiliary hypotheses any theoretical system can be immunised against falsificiations. However, the mere introduction of a new auxiliary hypothesis does not suffice in Lakatos's example in order to immunise the theory. Initial conditions and universal hypotheses also have to be changed. Therefore the conventionalist thesis should be formulated as saying that a theoretical system can always be immunised against criticism through appropriate modifications of it. If we modify the theoretical system in the way suggested by Lakatos and assume that the hidden magnet (or the unknown force) is strong enough, then we can no longer predict that the thread will break and thus save the theoretical system from falsification.

The mere introduction of a new auxiliary hypothesis is never sufficient in order to avoid a falsification. If a prognosis *P* follows from a theoretical system *T* and observations show that *P* is false, then the theory *T* is falsified. If we introduce the auxiliary hypothesis *AH*, then the theoretical system $(T \wedge AH)$ is still falsified because the prognosis *P* still follows from *T* and thus also from $(T \wedge AH)$. If $\neg P$ can be derived from *AH*, then the theoretical system $(T \wedge AH)$ is even inconsistent: *P* and $\neg P$ can be derived from it. Thus we have to modify *T* to *T´* in order to avoid falsification (and inconsistency) of the system $(T \wedge AH)$ (cf. chap. 7, B, below).

In Lakatos's example the following modifications are necessary in order to avoid the falsification of the theoretical system and in order to explain why the thread does not break:

> If and only if a thread is pulled by a weight exceeding its tensile strength, then it will break. (Hypotheses$_1$: H_1)
> The tensile strength of this thread is 1 *kp*. (Initial condition$_1$: C_1)
> The weight put on this thread in the space-time position *k* pulled it with a force of 2 *kp*. (Initial condition$_2$: C_2)
> In the ceiling of the laboratory there was a hidden magnet pulling the thread with a force of 1,5 *kp*. (Initial condition$_3$: C_3)
> ___
> The thread in the space-time position *k* does not break. (Prognosis: $\neg P$)

Thus extensive modifications of the theoretical system are necessary in order to explain why the thread did not break. The mere introduction of an auxiliary hypothesis about a hidden magnet does not suffice.

Lakatos discusses other modifications of the theoretical system. It can also be saved from falsification by modification of the first initial condition: "We propose that the tensile strength *does* depend on how moist threads are; the tensile strength of the actual thread, since it got moist, was 2 *lbs.*" Also the second initial condition can be modified: "the weight was only 1 *lb;* the scales went wrong." The truth value of *P* can be questioned: the observation that the thread breaks was wrong, because some "revolutionary laboratory assistants" regularly observed wrongly in order to annoy a liberal professor (Lakatos 1970, 185). Lakatos (1970, 186) concludes: "Indeed, there are infinitely many possibilities of how to replace—given sufficient imagination—any of the premises (*in the deductive model*) by invoking a change in some *distant* part of our total knowledge (*outside the deductive model*) and thereby restore consistency." Lakatos wants to reduce these "infinitely many possibilities" of modifying a theoretical system in sophisticated methodological falsificationism through the requirement that theoretical modifications should be "progressive," that is, that they should have an "excess" of empirical content.

In sophisticated methodological falsificationism fewer methodological (or conventional) decisions are necessary than in naive methodological falsificationism: the decisions about *ceteris paribus* clauses and metaphysical theories are superfluous. The conventional element in the decisions about test statements is at least reduced by the introduction of an appeal procedure (Lakatos 1970, 125–27).

Nevertheless, sophisticated methodological falsificationism also has many conventionalist traits. Like the conventionalist Le Roy, Lakatos maintains that science creates its own universe (1970, 188; Le Roy 1899–1900, 514–18; Poincaré 1970, 155). Although Lakatos adds that he uses 'create' in a "provocative-idiosyncratic sense," his conclusions are far from a criticist and falsificationist methodology. He (Lakatos 1970, 187) writes for example:

> The direction of science is determined primarily by human creative imagination and not by the universe of facts which surrounds us. Creative imagination is likely to find corroborating novel evidence even for the most 'absurd' programme, if the search has sufficient drive. This look-out for *new confirming evidence* is perfectly permissible. Scientists dream up phantasies and then pursue a highly selective hunt for new facts which fit these phantasies.

This search for supporting evidence is characteristic for inductivism. That Lakatos gives up (logical) falsificiations has its price and forces him to go back to inductivist and conventionalist ideas.

D. METHODOLOGY OF SCIENTIFIC RESEARCH PROGRAMMES

There are two problems in sophisticated methodological falsificationism which Lakatos hopes to solve in his methodology of scientific research programmes. The first problem is that it is too easy to find excess empirical content and progressive problem shifts in sophisticated methodological falsificationism. According to Lakatos's definitions, it is sufficient to add any completely disconnected low-level hypotheses to a theory in order to get excess empirical content. In order to avoid this "tacking paradox" Lakatos finds it necessary to demand that new additional hypotheses are more intimately connected with the original theory than by mere conjunction (Lakatos 1970, 131).

The second problem is the "remarkable continuity" of science, a continuity described by Kuhn as characteristic of "normal science." In sophisticated methodological falsificationism any part of a falsified theory may be changed, provided that these changes constitute a progressive problem shift and give excess empirical content. However, the continuity in the development of science seems to show that theories cannot be modified so freely, that some central parts of theories are kept unchanged. In order to explain this continuity of science, Kuhn introduced paradigms. Lakatos (1970, 132–33) introduces research programmes for the same reason.

In his *Postscript to the Logic of Scientific Discovery* Popper (Popper 1983, § 23) discusses metaphysical research programmes, for example the Cartesian research programme in physics, and says that they have a useful heuristic function. Lakatos introduces *scientific* research programmes and stresses their heuristic function. Furthermore, Lakatos thinks that they can explain the continuity of science. One of Lakatos's aims is to find a rational heuristic for science, to show that the new hypotheses and discoveries in science are the result of a continuous and rational development of thoughts (Duhem 1974, 220–52; Hanson 1958, 71; Lakatos 1974, 263, fn. 5).

In order to explain the continuity of science, Lakatos assumes that scientists usually do not abandon falsified theories, but modify them more or less extensively. In this context he finds it interesting that Popper allows that theories are defended tenaciously against criticism. According to Popper (1965, 312, fn. 1) "the dogmatic attitude of sticking to a theory as long as possible is of considerable significance." Without it "we shall never find out where the real power of our theories lies" (Popper 1970b, 55). Lakatos (Lakatos 1968b, 167, fn. 55) finds Popper's approval of a certain dogmatism puzzling and anomalous in falsificationism.

In order to understand Lakatos's ideas on research programmes, it is useful to discuss how falsified theories should be modified, whether they should be totally eliminated or only slightly modified. As Duhem (1974, 216–18) and Kuhn emphasise, falsified theories are not always totally eliminated. In order to explain this, Lakatos assumes that a research programme consists of two parts, a hard core of central hypotheses and a protective belt of auxiliary hypotheses. If a research programme is falsified, the protective belt of auxiliary hypotheses is modified, whereas the hard core of central hypotheses is unchanged:

> The negative heuristic of the programme forbids us to direct the *modus tollens* at this 'hard core'. Instead, we must use our ingenuity to articulate or even invent 'auxiliary hypotheses', which form a *protective belt* around this core, and we must redirect the *modus tollens* to *these* (Lakatos 1970, 133).

From a logical point of view, it is impossible to direct the *modus tollens* at some part of a falsified theoretical system. But we can *decide* to regard parts of a falsified theoretical system as unproblematic and hence not to modify them. Lakatos (1970, 135) expresses this by saying that a research programme has a hard core that is made irrefutable by "the methodological decision of its protagonists." In naive methodological falsificationism Lakatos thinks that test statements are made irrefutable by methodological decisions. In the methodology of scientific research programmes he uses the same idea in order to make the hard core of a research programme irrefutable. However, he also stresses that this hard core is the result of a long historical development: "The actual hard core of a programme does not actually emerge fully armed like Athena from the head of Zeus. It develops slowly, by a long, preliminary process of trial and error" (Lakatos 1970, 133, fn. 4). The relation between the preliminary process of trial and error and the methodological decision is unclear. Probably Lakatos thinks that the hard core of a research programme is decided to be "unrefutable" when the programme has been sufficiently tested.

The *positive heuristic* of a research programme consists of "a partially articulated set of suggestions or hints on how to change, develop the 'refutable variants' of the research-programme, how to modify, sophisticate, the 'refutable' protective belt." In this way the scientist is saved "from becoming confused by the ocean of anomalies" which Lakatos thinks surrounds every research programme (Lakatos 1970, 133). He does not think that falsifications are important: "Our considerations show that the positive heuristic forges ahead with almost complete disregard for 'refutations': it may seem that it is the *'verifications'* rather than the refutations which provide the contact points with reality" (Lakatos 1970, 137). In this way Lakatos hopes to explain the relative autonomy of theoretical science and the continuity of its historical development.

Lakatos wants to demarcate scientific research programmes from unscientific ones. In order to be scientific, a research programme must be "consistently content-increasing," that is: (a) each step in its development must lead to predictions of some new facts (consistently progressive theoretical shifts), (b) now and then some of these new predictions must be corroborated (intermittently progressive empirical shifts) (Lakatos 1970, 134). These two requirements for scientific research programmes are similar to the two requirements for a scientific theory in sophisticated methodological falsificationism. In both cases it is required (a) that the theory or programme leads to new empirical predictions (excess empirical content) (b) which now and then are corroborated, that is, some of the new predictions turn out to be correct. In the methodology of scientific research programmes we judge the development of a theory governed by a research programme and compare it to the development of another theory governed by another research programme.

Lakatos asks whether there are any objective criteria for the elimination of research programmes, in contradistinction to the subjective criteria for paradigms discussed by Kuhn. Lakatos's answer is, in outline, that "such an objective reason is provided by a rival research programme which explains the previous success of its rival and supersedes it by a further display of *heuristic power*." Lakatos uses 'heuristic power' to "characterize the power of a research programme to anticipate theoretically novel facts in its growth," that is, the power of a programme to generate excess empirical content, excess explanatory power (Lakatos 1970, 155, text and fn. 3).

How do we determine the heuristic power of a research programme? How do we compare it to the heuristic power of another programme? It is possible to determine if a research programme can explain the earlier success of a rival research programme. But how do we determine its future performances, its heuristic power? Lakatos does not answer this question, which shows the Achilles' heel of the methodology of scientific research programmes (cf. chap. 5, B 1, below).

The similarities between the elimination rules for research programmes and for scientific theories are obvious. In both cases it is essential that the successes of the rival programme or theory can be explained. With small modifications Lakatos transfers the elimination rule of sophisticated methodological falsificationism to the methodology of scientific research programmes. However, in the methodology of scientific research programmes new problems arise because its elimination rule is based on the future performance of the programme.

E. Rational Reconstruction of The History Of Science

In order to get rid of the annoying problem of how to judge the future performance and heuristic potential of a research programme, it is possible to modify the methodology of scientific research programmes by saying that only its past performances are relevant. Then only a rational reconstruction of the history of science is possible, not a methodology of scientific research programmes. In his later works Lakatos maintains that his aim is only to give rational reconstructions of the history of science, not to advise the scientist which research programme to work on (Lakatos 1971a; Lakatos 1971b, 174, 178; Lakatos, and Zahar 1975). This means that Lakatos does not want to give any methodological advice any longer, that he retires from the position of a methodologist in order to become a historian studying the problem shifts in the historical development of science and trying to reconstruct them rationally (cf. Musgrave 1976, 475–91).

The central problem in Lakatos's methodology of scientific research programmes is to answer the question of when a research programme should be eliminated. According to Lakatos (1970, 157, fn. 1) the pragmatic meaning of the rejection or elimination of a research programme is crystal clear: "it means *the decision to cease working on it*". When Lakatos later maintains that he does not want to advise the scientists which research programme to work on, then he has given up the elimination rule that is central for the methodology of scientific research programmes.

Is it possible to give a rational reconstruction of the history of science without a methodology? How do we determine what is rational in the history of science without methodological judgements and rules? In order to solve this problem, Lakatos introduces the accepted basic value judgement of the scientific élite (Lakatos 1978, 1: 131–36). He maintains that the scientific élite agrees on what is rational in science, at least on what particular single steps in the development of science are rational. Why are these judgements of the scientific élite rational? Who belongs to this élite? Already in order to answer these questions, normative and methodological judgements are necessary. Determining who belongs to the *scientific* élite is a methodological problem. According to a naturalistic conception of methodology rejected by Popper (1959, § 10), "methodology is an empirical study in its turn—a study of the actual behaviour of scientists, or of the actual procedure of 'science'." Popper thinks that it is a methodological problem to decide "what is to be called a 'science' and who is to be called a 'scientist'." These problems cannot be solved in a "naturalistic" way by mere descriptions. Therefore Popper rejects the naturalistic conception of methodology as uncritical. Popper's arguments against naturalism can also be used against Lakatos's rational reconstructions of the history of science.

From a historical point of view, Lakatos's rational reconstructions of the history of science are highly problematic. Although he (Lakatos 1970, 140, fn. 4) maintains that some of his reconstructions may "strike the historian as more a caricature than a sketch" and adds that they are "to be taken not with a grain, but with tons, of salt," the question remains concerning what deviations from the actual the history of science are allowed in rational reconstructions. In some cases it seems as if Lakatos does not reconstruct the history of science, neither rationally nor irrationally, but simply misunderstands it. For example, Lakatos tries to explain by a rational reconstruction why the Copernican theory was accepted so soon, why it was immediately superior to the Ptolemaic theory (Lakatos, and Zahar 1975, 360, 375–81). However, the history of science shows that the Copernican theory was not accepted immediately. On the contrary, the resistance against it was prolonged not only among the public, but also among scientists (Stimson 1917). This resistance was rational because the Copernican theory was confronted with serious problems, as, among others, Feyerabend has shown (cf. chap. 5, C 3, below).

Lakatos does not succeed in solving the problems raised by Kuhn. First Lakatos tries to formulate a normative methodology in order to save the rationality of science from relativism. When Lakatos's methodology of scientific research programmes is confronted with serious problems, he withdraws from methodology to a rational reconstruction of the history of science and hopes to answer at least some of Kuhn's questions. If a rational normative methodology is impossible, Lakatos wants to show that we can at least rationally reconstruct the history of science.

As was argued above, you cannot rationally reconstruct history without a normative methodology. Lakatos tries to solve the normative problems by describing the scientific élite and its basic value statements. This methodological naturalism is not convincing. Lakatos's attempt to answer the methodological problems raised by the history of science ends in the view that the history of science, personalised in the scientific élite, can be enthroned as the supreme judge of methodological controversies. Thus Lakatos's attempt to save the rationality of science ends in a defeat, total and complete.

CHAPTER FIVE

FEYERABEND'S EPISTEMOLOGICAL ANARCHISM

Feyerabend belongs, together with Kuhn, to a group of philosophers who criticise the positivist and the falsificationist views of science. But he also criticises Kuhn's and Lakatos's views of science. Feyerabend rejects Kuhn's views of normal science because it does not agree with the history of science and because it has dogmatic consequences. Against Lakatos and Popper he objects that they do not succeed in showing that science is rational. In his criticism of critical rationalism Feyerabend has found followers among the contemporary critics of science and technology.

Feyerabend (1975) first calls his position epistemological anarchism. He regards science as fallible and experience as theory-dependent. So far his views are in agreement with critical rationalism. In contradistinction to it, Feyerabend does not think that a critical and rational discussion of competing theories is possible. Like other followers of the new philosophy of science, Feyerabend thinks that the theory-dependence of experience makes competing theories incommensurable.

Feyerabend's views of scientific theories are influenced by Wittgenstein's views of different "forms of life" (Bartley 1973, 170–71; Wittgenstein 1967, § 23, 241). In contradistinction to Wittgenstein, Feyerabend wants many competing forms life, pluralism of theories and of forms of life. According to Feyerabend we can only criticise a theory or a form of life from the point of view of another theory or form of life. Since he thinks that theories and forms of life are incommensurable, there is no common basis for such critical discussions, and theories and forms of life have to be discussed in an anarchistic way by using rhetoric and propaganda. In order to show this Feyerabend (1988, chaps. 6–10, 13, 14) discusses Galileo's defence of the Copernican theory.

Feyerabend does not think that Lakatos succeeded in saving reason in science. The first edition of *Against Method* was dedicated to Lakatos as friend and fellow-anarchist. There Feyerabend argues that Lakatos's methodology of scientific research programmes fails and that Lakatos is an epistemological anarchist in disguise.

A. Criticism of Kuhn

Feyerabend finds Kuhn's presentation of science ambiguous and asks whether Kuhn wants to give methodological prescriptions or historical descriptions, whether he wants to tell the scientists how to proceed or whether he wants to describe the actual development of science (Feyerabend 1970, 198).

First Feyerabend assumes that Kuhn wants to describe the actual development of science. According to Kuhn, science is set apart from other activities by a tradition of "puzzle-solving." Feyerabend (1970, 199–200) does not think that such a tradition is typical for science, since there are many other puzzle-solving traditions, for example "Oxford philosophy, or, to take an even more extreme example, *organized crime*." In addition, and this is more important, Feyerabend asks whether normal science, as described by Kuhn, is a necessary condition of scientific revolutions. According to Kuhn, science is governed by a paradigm in normal science. Feyerabend thinks that science needs a theoretical perspective. "To this common idea Kuhn adds a specific twist of his own. He defends not only the *use* of theoretical assumptions, but the *exclusive choice* of one particular set of ideas, the monomaniac concern with only one single point of view" (Feyerabend 1970, 201). Feyerabend does not think that this is a correct description of the actual development of science. He thinks that the history of science is a history of competing theories, that a theory is criticised from the point of view of another theory not only when scientific revolutions occur, but also in normal science. Thus Feyerabend finds Kuhn's description of normal science false.

Feyerabend (1970, 206) does not think that Kuhn can explain why scientific revolutions occur or why new theories arise: "If normal science is *de facto* as monolithic as Kuhn makes it out to be, then where do the competing theories come from?" According to Kuhn there has to be a crisis in order to motivate scientists to suggest new theories. In his conception a crisis is a precondition for a competing theory. Feyerabend thinks that the opposite is true: only when there is a competing theory do scientists discover the weak points of an established theory; only a competing theory can cause a crisis of an established theory. If problems are regarded as puzzles in normal science, why does the tradition of normal science not continue forever, why are all new problems not regarded as puzzles or anomalies to be solved by the established paradigm? Feyerabend does not think that Kuhn can explain why crises and revolutions appear in the development of science.

In order to show that a theoretical alternative is a necessary condition for a scientific revolution and for a crisis, Feyerabend discusses the

Copernican revolution. He denies that the Ptolemaic theory was in a crisis or threatened by serious anomalies when Copernicus suggested his new theory. On the contrary, Copernicus did not present any new facts that would refute the old theory, not even new facts that could not be explained by it (Copernicus 1939, 57; Feyerabend 1978, 223). First the Copernican theory led the scientists to question the Ptolemaic theory and to take its problems seriously. The weak points of a theory are discovered first when there is a competing theory (Feyerabend 1970, 206–7).

Feyerabend argues that Kuhn does not describe the development of science correctly. There is no "normal science" with only one dominant theory, but the history of science is the history of competing theories. Crises do not lead to "extraordinary science" and the search for competing theories. On the contrary, crises and scientific revolutions are the results of competing theories. In the history of science there is an active interaction between competing theories. Feyerabend (1970, 202) concludes that "Kuhn's account which *temporally separates* periods of proliferation [of theories] and periods of monism altogether collapses." This is an important historical criticism of Kuhn's model for the development of science. It shows that Kuhn's ideas on the structure of scientific revolutions are problematic from a historical point of view. Feyerabend argues that Kuhn's main example of a scientific revolution, the Copernican revolution, was not caused by a crisis. He also argues that Kuhn's description of dogmatic normal science is incorrect and depends on the fact that Kuhn exaggerates the guiding role of theories in scientific research. Theoretical presuppositions do not have to be understood as dogmas excluding any theoretical alternative, as in Kuhn's description of normal science.

Feyerabend does not find Kuhn's implicit methodological recommendations acceptable. If our aim is scientific revolutions and changes of paradigms, "then we must be prepared to introduce and articulate alternatives to [the ruling theory] *T* or, as we shall express it ..., we must be prepared to accept a *principle of proliferation*" (Feyerabend 1970, 205). According to this principle we must always be prepared to introduce competing theories in order to make theoretical pluralism, development and scientific revolutions possible.

At the same time Feyerabend (1970, 203–4) recommends that theories should be tenaciously defended (principle of tenacity). He does not suggest that all scientists should defend the same theory (or paradigm), as in Kuhn's normal science, but that different groups of scientists should defend different theories. In this way we would get many competing theories, that is theoretical pluralism. The principles of proliferation and tenacity should not govern different phases of research, as Kuhn recommends, but should operate at the same time in order that there is always competition between tenaciously defended theories. Feyerabend recommends theoretical pluralism.

Why does Feyerabend think that theoretical pluralism is good? For example, does it facilitate the growth of knowledge, as a critical rationalist would say? This answer is not available to Feyerabend because he thinks, like Kuhn, that competing theories are incommensurable. For this reason Feyerabend cannot argue that the competition of theories leads to progress or growth of knowledge (cf. Feyerabend 1970, 202). Since competing theories are incommensurable, we cannot say that there is theoretical progress, only that there is theoretical change.

With the idea of growth of knowledge Feyerabend also rejects phrases such as "search for truth" or "highest aim of mankind." Although such phrases are liberally used in order to ennoble science, Feyerabend (1970, 209) thinks that they at the same time remove science from the domain of critical discussion. Science is only one of many possible forms of life, according to Feyerabend. He does not think that just this form of life should be uncritically accepted and used as a measure for the rationality of other forms of life. Here Feyerabend uses his principle of proliferation on a higher level in order to discuss different forms of life:

> Progress has always been achieved by probing well-entrenched and well-founded forms of life with unpopular and unfounded values. This is how man gradually freed himself from fear and from the tyranny of unexamined systems (Feyerabend 1970, 202–3).

As different forms of life are incommensurable, it is unclear what Feyerabend means by progress. As an epistemological anarchist he can hardly mean growth of knowledge.

Since Feyerabend cannot argue that theoretical pluralism facilitates growth of knowledge, he introduces the new idea that it contributes to the happiness and full development of human beings. Thus he abandons the argument that theoretical pluralism leads to the development of knowledge and instead uses the argument that it leads to the development of human beings and increases happiness. This, without doubt, is also a kind of progress. A tenacious defender of the principle of incommensurability could argue that the happiness and development of human beings are also incommensurable. Feyerabend (1970, 210) does not discuss this problem, but thinks that an undogmatic attitude contributes to happiness and human development and exclaims: "But this is a happy coincidence indeed!"

Feyerabend (1970, 210) also introduces a biological argument for pluralism:

> The interplay between proliferation and tenacity also amounts to the continuation, on a new level, of the biological development of the species and it may even increase the tendency for useful *biological* mutations. It may be the only possible means of preventing our species from stagnation.

After Feyerabend has abandoned the idea of growth of knowledge, he must search for ethical and biological arguments for his methodological recommendations. It is not so easy for Feyerabend to motivate his own methodological proposals. This does not invalidate his criticism of Kuhn's conceptions. Kuhn's model for the development of science is, as Feyerabend argues, historically inadequate. And if we are interested in development and change of knowledge, then Kuhn's implicit methodological recommendations are unacceptable, independent of whether our ultimate aim is growth of knowledge, happiness, human perfection, or biological development.

B. CRITICISM OF LAKATOS

1. *Criticism of the Methodology of Scientific Research Programmes*

Feyerabend doubts that it is rational to eliminate degenerative research programmes, as Lakatos recommends in his methodology of scientific research programmes. In his discussion of naive methodological falsificationism, Lakatos says that falsified theories should not be eliminated because they may develop. Feyerabend (1975, 185) asks whether this is not also true for research programmes: "If it is unwise to reject faulty theories the moment they are born because they might grow and improve, then it is also unwise to reject research programmes on a downward trend because they might recover and might attain unforeseen splendour (the butterfly emerges when the caterpillar has reached its lowest stage of degeneration)."

Lakatos's rule for the elimination of research programmes would be rational if research programmes that were degenerative in the past continued to be degenerative in the future. However, this presupposes an inductive principle telling us that the future will be similar to the past, and hence "Lakatos is ... and ... cannot help being, a second-level inductivist" (Feigl 1971, 147).

To avoid this accusation, Lakatos (1971b, 174) has maintained that he does not recommend the elimination of degenerative research programmes. According to Feyerabend this means that in practice Lakatos has adopted a position of epistemological anarchism according to which "everything goes." The methodology of scientific research programmes "provides *standards* that aid the scientist in evaluating the historical situation in which he makes his decisions; it does not contain *rules* that tell him what to do" (Feyerabend 1975, 186). A scientist may eliminate a degenerative research programme. Since Lakatos does not want to give any methodological advice, the scientist is totally entitled to do the opposite. Degenerative programmes "can always stage a comeback" (Lakatos 1971b, 174).

Earlier Lakatos did not make so far-reaching concessions, but only recommended that his elimination rule for research programmes should be used in the long run and that young research programmes should be given a breathing space. How long should this breathing space be? Since Lakatos does not give any time limits, Feyerabend (1970, 215) finds that the idea of a breathing space is already anarchism in disguise.

Feyerabend thinks that Lakatos keeps his methodological standards merely as a "*verbal ornament*, as a memorial to happier times when it was still thought possible to run a complex and often catastrophic business like science by following a few simple and 'rational' rules." He confronts Lakatos with the following dilemma: the methodology of scientific research programmes is either a normative methodology governing research or an expression of epistemological anarchism. The first alternative leads to the problem of reconciling methodological advice with the history of science; the second to the original problem of relativism, which Lakatos wanted to solve.

Musgrave (1976, 208) has formulated this dilemma in the following way: Lakatos has to choose between inductivism and relativism. In order to give methodological advice, Lakatos has to make inductivist assumptions. Without such advice Lakatos is in fact a relativist. Both alternatives are unacceptable to Lakatos. His principal aim was to overcome the relativist consequences of the study of the history of science. Without a normative methodology he has failed to reach this aim. To make inductivist assumptions in order to overcome relativism cannot be acceptable to anybody cognisant of Popper's severe criticism of inductivism. As Feigl (1971, 147) put it: "Sir Karl has put 'the fear of God' into (most) inductivists." Although Feigl tries to defend a kind of inductivism, Lakatos does certainly not want to do so.

Musgrave tries to save Lakatos from the dilemma by changing Lakatos's rule of elimination so that not all, but most scientists should stop working on a degenerative programme: "An appraisal like 'This programme has degenerated until till now, while its rival has progressed' can and should lead to heuristic advice of the following kind: 'Science should devote most of its resources to the progressive programme.'" In this way Musgrave hopes to save Lakatos from relativism.

Does Musgrave save Lakatos from inductivism? The methodological advice that most scientists should work on progressive research programmes is problematic: "At this point those with a keen nose for such things will smell an inductivist rat" (Musgrave 1976, 480). The tacit presupposition that progressive research programmes will continue to be so is crude inductivism and sinful enough on any level.

> But on Feigl's 'second level', where it concerns the essentially unpredictable growth of knowledge, it is a mortal sin of the first order. Have we rescued Lakatos's methodology from the frying-pan of anarchism, only to see it cast into the hell-fire of inductivism (Musgrave 1976, 480)?

Musgrave does not think so because Lakatos's definition of a progressive research programme can be changed in the following way: "A progressing research programme throws up more unsolved but solvable problems than a degenerating one" (Musgrave 1976, 482). Is Lakatos rescued from inductivism in this way? That a research programme throws up more unsolved but solvable problems than another can only refer to past achievements of the programme. There cannot be any guarantee that the research programme will continue to do so in the future. Perhaps the recommendation that science ought to devote more energy to progressing than to degenerating programmes can be upheld nevertheless. Musgrave (1976, 482) does not think that a guarantee is necessary, but only conjectures that progressing research programmes will continue to be progressive: "Of course, there is no guarantee that following such advice will lead to success." Musgrave's new definition raises another problem. How can you know that a problem is solvable before you have found a solution? Musgrave (1976, 482) says that a "progressive research programme is throwing up a rich store of interesting unsolved problems, and hints about how to solve them." These hints probably are the basis for conjectures about the solvability of problems.

Musgrave modifies Lakatos's methodology of scientific research programmes extensively in order to rescue it from inductivism. The modified version is compatible with a falsificationist conception of science and in no way an alternative, but a heuristic complement to it. Feyerabend (1976, 123) argues against Musgrave's modification: "Why should a research programme that creates lots of problems be preferable to a research programme that creates none?" This argument is rather weak since Musgrave speaks about solvable problems. But Feyerabend's arguments against Lakatos's original conception of a methodology of scientific research programmes are important. Feyerabend argues that Lakatos's methodology does not recommend or forbid anything and hence is relativism (or anarchism) in disguise, "a splendid Trojan horse" that can be used to smuggle relativism "into the minds of our most dedicated rationalists" (Feyerabend 1975, 200).

2. *Criticism of Lakatos's Rational Reconstructions of the History of Science*

In his last papers Lakatos tried to show how the history of science can be rationally reconstructed. In order to judge whether a reconstruction is *rational*, Lakatos (1971a, 111) compares it to the basic value judgements of the scientific élite.

Feyerabend objects that these judgements are seldom uniform. He argues that the history of science records that different famous scientists have often judged very differently. In addition, these basic value judgements were only rarely made for good reasons. Hence Feyerabend (1975, 203) concludes that "'*common scientific wisdom' is not very common and it certainly is not very wise.*" Feyerabend argues that the common scientific wisdom reflected in basic value judgements of the scientific élite only reflects the professional ideology of scientists. He asks why the scientific form of life should be regarded as the ultimate measure of rationality. This authority is arbitrary, especially if we, as Lakatos suggests, take the science of the last two centuries as our point of departure (Feyerabend 1981a, 2: 209). A description of the basic value judgements of the scientific élite does not give us criteria allowing us to judge the rationality of reconstructions of the history of science, but a description of the modern scientific form of life: "It is a topic neither for methodology, nor for the theory of rationality, but for 'mob psychology', ... or, to use a more traditional term, *it is a topic for the sociology of knowledge*" (Feyerabend 1976, 126).

C. CRITICISM OF POPPER

1. *Principle of Tenacity*

Feyerabend (1970, 203) suggests a principle of tenacity according to which scientists should stick to theories in spite of considerable difficulties. An example of such difficulties would be falsificiations, as Feyerabend argues in his historical case studies. Feyerabend introduces this principle in order to justify a research strategy typical for what Kuhn calls normal science, the strategy of not eliminating falsified theories: "Having adopted tenacity we can no longer use recalcitrant facts for removing a theory ... even if the facts should happen to be as plain and straight-forward as daylight itself" (Feyerabend 1970, 205).

Feyerabend thinks that the principle of tenacity is reasonable, because theories can be developed and improved, test statements can be rejected, and auxiliary hypotheses changed. By modifying and developing a theory you might improve it so much that earlier problems are solved. Observational or experimental errors are possible, and it takes time to get uniform experimental results. Implicit auxiliary hypotheses are used in order to derive test statements; explicit auxiliary hypotheses are used in order to decide what test statements to accept and what concepts to use in order to describe observations (Feyerabend 1970, 204–5).

Feyerabend uses historical cases to show that the tenacious defence of a theory can be rational. He thinks that in our world a falsificationist methodology meets insurmountable difficulties. Even though most theories are falsified, it is nevertheless rational to continue to work on them and to develop them. In this way Feyerabend tries to show that falsificationism is a methodology that cannot be used in actual research (Feyerabend 1978, 227).

2. *Incommensurability*

Feyerabend (1970, 219) wholeheartedly accepts Kuhn's idea that competing theories are incommensurable. He (1970, 223) thinks that it makes realism "an impossible doctrine" and regards it as a decisive argument against critical rationalism (Feyerabend 1978, 28, fn. *).

Incommensurable theories can be falsified by using their own kind of experience according to Feyerabend. Without a commensurable alternative, however, these refutations are quite weak. It is impossible to compare the empirical contents or the verisimilitude of theories except within the confines of a particular theory. Hence Popper's methods for showing that there is progress in science cannot be used. The method of refutation, which can be used, is very much reduced in strength. "What remains are aesthetic judgements, judgements of taste, and our own subjective wishes. Does this mean that we are ending up in subjectivism" (Feyerabend 1970, 228)? Feyerabend accepts subjectivism and tries to show that it is not at all unattractive. In a conventionalist way he argues that science, after all, is our own creation and that it is only one of many possible forms of life. "It is good to be constantly reminded of the fact that science as we know it today is not inescapable and that we may construct a world in which it plays no role whatever (such a world, I venture to suggest, would be more pleasant than the world we live in today)" (Feyerabend 1970, 228).

Feyerabend does not tell us much about this different and more pleasant world. He informs us that the choice of basic cosmology may become a matter of taste: "There are many ways open to us once the fact of incommensurability is understood, and taken seriously" (Feyerabend 1970, 229). The development to subjectivism "far from being undesirable, changes science from a stern and demanding mistress into an attractive and yielding courtesan who tries to anticipate every wish of her lover" (Feyerabend 1970, 229). Thus Feyerabend thinks that subjectivism leads to another, gayer science. How can Feyerabend be so sure that his own subjective values will determine the new science? If subjectivism is taken seriously and anything goes, what prevents that the subjective values of some big brother or powerful politician determining the new and alternative science? Bertrand Russell (1940, 148) argues that subjectivism and relativism have the consequence that "empirical truth can be determined by the police." Feyerabend's belief in the felicitous consequences of subjectivism is naive and unfounded: instead of a new and gay science, the consequence can very well be the end of science.

3. *Cases from the History of Science*

Feyerabend (1983, 383) complains that only few readers have studied his historical case studies carefully. They are of great methodological interest. Feyerabend tries to show that Galileo defended the Copernican theory in an anarchist way. In the beginning the Copernican theory was confronted with serious problems, like the problems of explaining the variations in size and brightness of the planets when observed with the naked eye. These observations contradicted the Copernican theory, but Galileo immunised them *ad hoc* by introducing the auxiliary hypotheses that observations with the naked eye are unreliable. Some physical experiments with freely falling bodies, like the tower experiment, also refuted the Copernican theory according to Feyerabend. Galileo also solved these problems *ad hoc* by introducing a new theory of motion. Galileo's assumption that the telescope is reliable and his introduction of a new theory of motion are used by Feyerabend in order to show that Galileo did not take falsifications seriously and that he immunised them by introducing *ad-hoc*-hypotheses.

3.1. *Galileo's Use of the Telescope*

Before the invention of the telescope, observations with the naked eye seemed to falsify the Copernican (and the Ptolemaic) theory of the planetary system. In both theories it is assumed that the distances between Earth and the planets vary. (In the geocentric Ptolemaic theory these variations are mainly due to the epicycles.) With the naked eye you can observe variations in size and brightness of the planets, but they are smaller than predicted by the two theories. Do these deviations between observation and prediction falsify both theories? Do they show that the theories are only instruments useful in order to predict the positions of the planets? In his anonymous preface to Copernicus' main work, Andreas Osiander argues in this way: the predicted sizes of Venus contradict the experience of all ages (Feyerabend 1975, 111). According to Galileo (1967, 334) Copernicus says nothing about the almost constant size of Venus when observed with the naked eye and about the variations in size of Mars he finds that Copernicus says much less than necessary, probably because he could not explain these variations satisfactorily. Does this show that Galileo acted according to Feyerabend's principle of tenacity and did not take falsifications seriously?

When Galileo observed the planets with the telescope, at that time a recently invented instrument, he made astonishing discoveries. With the telescope the variations in size of the planets could be observed much more accurately than with the naked eye. The observations with the telescope agreed with the Copernican theory. According to Feyerabend (1988, 108) this proved for Galileo that the Copernican theory is true and that the telescope is reliable.

Without inductivist assumptions no observations can prove the Copernican theory or the reliability of the telescope. Feyerabend argues that the telescope was a rather problematic instrument in the beginning. On the basis of his observations of the moon with the telescope Galileo made drawings of the moon that were unreliable. Galileo's first observations of the moons of Jupiter were problematic and could in the beginning not be repeated by some scientists. With the telescope Galileo got other values for the variation in size and brightness of the planets than with the naked eye. However, when observations with the telescope and the naked eye contradict each other, why assume that observations with the telescope are reliable? Why not, like Galileo's Aristotelian adversaries, assume that only observations with the naked eye are reliable (Feyerabend 1988, chaps. 8–9)? Feyerabend (1988, 109) thinks that the Copernican theory was confronted with serious problems, "but that being in harmony *with still further inadequate theories* it gained strength, and was retained, the refutations being made ineffective by *ad hoc* hypotheses and clever techniques of persuasion." Such "irrational" methods of support are needed because of the uneven development of the Copernican theory and its auxiliary hypotheses. With older auxiliary hypotheses, as for example the Aristotelian theory of motion or the hypotheses that observations with the naked eye are reliable, the Copernican theory could be refuted. The new auxiliary hypotheses about the telescope and motion were developed after the Copernican theory. Feyerabend (1988, chap. 11) regards such later introduced auxiliary hypotheses as *ad hoc* and as immunising criticism.

When in 1610 Galileo published the first account of his observations with the telescope in *Sidereus Nuncius*, the whole academic world turned against him and said that it is well known that instruments with lenses are unreliable and produce artefacts (Ronchi 1966, 61). Even if Galileo knew the optical theories of his time, as Stillman Drake (1970, 142) argues, it was impossible to construct a telescope with the help of them (Van Helden 1974, 40).

One of the few scientists to take Galileo's observations seriously right from the beginning was Johannes Kepler, who discusses them in *Dissertatio cum Nuncio Sidereo* published in 1610. Kepler thinks that the use of the telescope is justified only when the instrument can be theoretically explained and immediately tries to give such an explanation (Dijksterhuis 1961, 388). In *Dioptrice*, published in 1611, he succeeds in explaining Galileo's telescope and even succeeds in predicting how to construct a new and more efficient one (Dijksterhuis 1961, 389; Van Helden 1974, 41–42). It was later constructed and widely used for astronomical observations.

Feyerabend denies that Kepler explained how the telescope functioned only one year after the publication of Galileo's first telescopic observations, because he thinks that Kepler's optical theory was falsified. If you look at an object close to the focus of a magnifying glass, it should be seen very far away according to Kepler's theory. However, such a phenomenon cannot be observed (Feyerabend 1988, 104).

Feyerabend thinks that Galileo's observations with the telescope were problematic for the following reasons: (1) the observations with the telescope contradicted the observations with the naked eye, for example the observations of the size and brightness of the planets, (2) the observations of the moons of Jupiter could not be repeated by some observers in the beginning, (3) Galileo could not explain how the telescope functioned, (4) Kepler's explanation of the telescope used a theory that was easy to falsify. In this way Feyerabend wants to show that Galileo defended the use of the telescope tenaciously in spite of considerable difficulties.

3.2. *Galileo's New Theory of Motion*

In order to explain the observed diurnal motion of the sun (and the stars) Copernicus assumed that the earth rotates diurnally around its axis. Ptolemy had used a number of arguments against this hypothesis, which for centuries were regarded as conclusive, as for example: "If the earth really rotated round its axis, in the eastward direction—which would be necessary to explain the westward motion of the celestial sphere—a stone thrown up vertically would have to fall down west of the starting-point, for during the upward and downward movements of the projectile the earth has rotated farther in the eastward direction" (Dijksterhuis 1961,

63–64). Feyerabend discusses a similar argument used by Galileo against the hypothesis of the diurnal motion of the earth: A stone falling from a tower falls vertically and does not hit the earth far west away from the tower, as you would expect on the basis of the Aristotelian theory of motion and the hypothesis of the diurnal motion of the earth (Feyerabend 1988, chaps. 6–7). Like Ptolemy Tycho Brahe also regarded the tower experiment and similar experiments as falsifications of the Copernican theory (Dreyer 1953, 355–56). This falsification depends on the Aristotelian theory of motion. Dijksterhuis (1961, 63) regards it as *ad hoc.* Although the falsification was later regarded as unconvincing, the reason is not that it was *ad hoc*, but that it depends on the Aristotelian theory of motion, an auxiliary hypothesis that was later questioned.

Galileo replaces the Aristotelian theory of motion with a new theory, which also differs from the medieval impetus theory. With Galileo's new theory of motion it follows that stones fall vertically from a tower, independently of whether the earth moves or not (Dijksterhuis 1961, 333–59).

In the first edition of *Against Method* (chap. 8) Feyerabend argues that the Galilean theory of motion was *ad hoc* and introduced subsequently in order to rescue the Copernican theory from falsifications. This is a problematic idea (Andersson 1981a; Andersson 1991a (English translation)). In the revised edition of *Against method* Feyerabend leaves out this chapter and in some places weakens his criticism of falsificationism considerably:

> The idea of the motion of the earth was in big trouble or, if you will, 'refuted'. It could survive only if it was treated with kindness. But if *it* could be treated with kindness, then so could the older system (Feyerabend 1988, 150).

In spite of this, Feyerabend still maintains that falsificationism is untenable: "wherever we look, whatever examples we consider, we see that the principles of critical rationalism (take falsifications seriously; increase content; avoid *ad hoc* hypotheses; ...) ... give an inadequate account of the past development of science and are liable to hinder science in the future" (Feyerabend 1988, 163–64). In the revised edition Feyerabend says that it is misguided "to try reducing the process 'Copernican Revolution' to a single principle, such as the principle of falsification." But elsewhere Feyerabend still maintains (as in the first edition) that critical rationalism is not only misguided, but bound to wipe out science (Feyerabend 1988, 164).

Feyerabend thinks that Galileo introduced a new "natural interpretation" of the tower experiment. According to the Aristotelian interpretation of the experiment, every motion needs a moving force, every motion is operative. With this natural interpretation the tower experiment proves that the earth is at rest. According to the Galilean interpretation, only change of motion needs a moving force, that is only relative motion is operative. With this new interpretation the tower experiment does not prove that the earth is at rest (Feyerabend 1988, 72). Feyerabend also maintains that Galileo's new theory of motion led to a change of experience (Feyerabend 1988, 74). How is the experience of the tower experiment changed by Galileo's new dynamics? Galileo and his Aristotelian adversaries experience the tower experiment in the same way: the stone falls vertically down. What has changed is not the experience, but the auxiliary hypotheses used to predict the outcome of the tower experiment. Thus the change is not empirical, but theoretical, a change from the Aristotelian to the Galilean theory of motion. Problematic is only why Feyerabend maintains that this theory introduced a new and highly speculative experience (Feyerabend 1988, 78). Neither the Aristotelian nor the Galilean theory of motion is a natural interpretation of experience, but a speculative and theoretical explanation of what has been observed.

D. Epistemological Anarchism and Critical Rationalism

According to Bartley, the question of how forms of life and language games change is one of the most difficult problems in Wittgenstein's late philosophy. Wittgenstein thinks that forms of life change and develop. However, he insists that one form of life is not allowed to judge another and that philosophy must not play any role in this process (Bartley 1973, 170–71; Wittgenstein 1967, § 124). Feyerabend wants changes of scientific theories and thinks that an undogmatic, open attitude contributes to human happiness and development. But he does not think that a theory can be criticised by an independent experience (thesis of incommensurability). Hence he thinks rhetoric, propaganda, and competing theories are necessary in order to criticise and develop other scientific theories. Thus Feyerabend's philosophy of science is an attempt to solve a basic problem in Wittgenstein's late philosophy. According to Feyerabend, a theory can be criticised only by another theory. For this reason theory pluralism plays a different role in Feyerabend's epistemological anarchism than in critical rationalism, which also finds theory pluralism desirable, but not absolutely necessary for criticism.

According to critical rationalism, we can always find a common empirical basis when we compare different theories. Hence Popper regards the thesis of incommensurability is an expression of "the myth of the framework" (Popper 1970b, 56). Of less importance are the differences related to the role of auxiliary hypotheses in the development and testing of scientific theories. I will investigate these methodological differences in the following chapters.

However, I will not try to answer Feyerabend's (1976, 110) question: "What's so great about science"? I will not appraise science and compare it with other forms of life. From the point of view of critical rationalism, science and especially uncritical belief in science can also be criticised. Therefore, I follow Feyerabend's fight against scientism with sympathy, even if I do not think that we have to jump from scientism into postmodern relativism.

CHAPTER SIX

A CRITICIST SOLUTION TO THE PROBLEM OF THE EMPIRICAL BASIS

To be able to test scientific theories empirically, you need singular statements about observations and experiments. For historical reasons Popper calls these statements basic statements in *The Logic of Scientific Discovery*. But he stresses that these statements are fallible and revisable and do not belong to any absolute empirical basis. Later Popper has added that the word 'basis' has ironical overtones: "it is a basis that is *not firm*. ... Our observational experiences are never beyond being tested; and they are impregnated with theories" (Popper 1959, addendum, 1972, 111). Nevertheless, Popper thinks that basic statements are intersubjectively testable and in this sense objective. Since they are fallible and used to test theories, I will use the term 'test statements' instead of 'basic statements'.

Kuhn, Lakatos, and Feyerabend also think that test statements are fallible and revisable and that they depend on theories. Like Popper, they criticise the positivist assumption that science has an infallible empirical basis. On this point Popper and his critics agree. However, Popper's critics accuse him for not having solved the methodological problems arising when test statements are fallible.

Kuhn asks how theories can be falsified with fallible test statements. Like Feyerabend, he thinks that competing theories are incommensurable because test statements depend on theories. Feyerabend motivates his relativism and scepticism with the theory-dependence of experience. He thinks that only methods like rhetoric and propaganda can be use, if competing theories are incommensurable. Lakatos does not think that the method recommended by Popper for criticism of test statements, derivation of further consequences, can be used in practice. According to Lakatos we do not criticise test statements directly, but indirectly through their interpretative theories.

In this chapter I will investigate whether Popper's solution of the problem of the "empirical basis" is satisfactory and whether it answers the questions asked by Kuhn, Lakatos, and Feyerabend.

A. Popper on Test Statements

1. *Function of Test Statements*

The most important function of test statements is to test scientific theories. We need them to decide whether a theory is falsifiable (scientific) and in order to corroborate or falsify it (Popper 1959, § 28). In order to falsify a theory, we need not only test statements, but also "falsifying hypotheses" to guarantee that the falsification is reproducible (Popper 1959, § 22).

However, test statements are used not only in order to discuss scientific theories critically, but also in order to discuss other test statements. The test statements themselves are intersubjectively testable and objectively criticisable (Popper 1959, 104, 111). It is essential for Popper's solution of the problem of the "empirical basis" that test statements themselves can be critically discussed. Test statements are used not only in order to test theories, but also in order to test other test statements.

2. *Logical Form of Test Statements*

With the help of test statements it is possible to falsify scientific theories. Test statements are potential falsifiers of a scientific theory. For example, a potential falsifier to the general hypotheses 'All swans are white' is 'There is a swan that is not white here now' or 'In the time-space-region *k* there is a non-white swan'. Popper calls such existential statements singular existential statements. They have the logical form 'In the time-space-region *k* there is a *P*'. As a formal requirement Popper (1959, § 28) lays down that all test statements should be singular existential statements.

This requirement leads to problems. The first type of falsification dealt with by Popper in *Logic of Scientific Discovery* is the falsification of a theoretical system consisting of general hypotheses and singular initial conditions (see above chap. 2, C1). From these systems we can derive (unconditional) prognoses, like 'In the space-time-region *k* there is a white swan' and 'In the time-space-region *k* there is a thread that does not break'. Such systems can be falsified by the negations of prognoses, that is by singular non-existence statements (having the logical form 'In the space-time-region *k* there is no *P*'), but they cannot be falsified by any singular existential statement. Thus Popper's formal requirement that test statements should be singular existential statements is too narrow. Potential falsifiers of theoretical systems consisting of general hypotheses and singular initial conditions do not meet this requirement.

Popper's formal requirement for test statements leads to a second problem. According to Popper, test statements are used not only in order to test general hypotheses, but also in order to test other test statements. It should be possible to falsify a test statement with another test statement. According to Popper's criterion of demarcation, scientific statements should be falsifiable. Thus we require that test statements should be falsifiable, that is that it should be possible to falsify them with other test statements. However, if test statements are singular existential statements, they cannot be falsified by other test statements. Hence test statements cannot be criticised by other test statements. They are not even scientific (in the sense of being falsifiable) (Kneale 1974, 206–8). The reason for this second problem is that no singular existential statement can contradict another singular existential statement. There is for example no contradiction between the statements 'In the time-space-region *k* there is a *P*' and 'In the time-space-region *k* there is a non-*P*'.

Popper finds the consequence that test statements are unscientific "outrageously silly." In an answer to the criticism presented above, he writes that test statements are singular statements, "and since two singular statements can clash with each other, singular test statements (that is, 'basic statements') ... fall ... within the province of empirical science, according to my criterion of demarcation" (Popper 1974, 988). It is correct that two purely singular statements of the type 'The raven *a* is white' and 'The raven *a* is not white' do contradict each other. But it is not correct for existential statements, not even for singular existential statements of the type 'In the space-time-region *k* there is a *P*'. There is no logical contradiction between 'In the space-time-region *k* there is a white swan' and 'In the space-time-region *k* there is a non-white swan'. (From a logical point of view there might very well be two swans in the space-time-region, a white, and a non-white one.) Popper explicitly demands that test statements should be singular existential statements. He even writes that "a basic statement must have a logical form such that its negation cannot be a basic statement in its turn" (Popper 1959, 101). Thus Popper's formal requirement has the consequence that test statements are unfalsifiable and hence unscientific.

This consequence is unintended. Popper (1959, § 6) maintains that the test statement 'It will rain here tomorrow' is scientific. It is contradicted by the statement 'It will not rain here tomorrow.' This statement can also be expressed: 'In the space-time-region *k* there is no rain.' Thus it is a singular non-existence statement, and according to Popper's formal requirement no test statement. For this reason the statement 'It will rain here tomorrow' is unfalsifiable, since it cannot be refuted by any singular existential statement. Popper writes: "Strictly existential statements ... cannot be falsified. No singular statement ... can contradict the existential statement 'There are white ravens'" (Popper 1959, § 15, 69). This is true not only for strictly existential statements, but also for singular existential statements. No singular statement can contradict the singular existential statement: 'In the time-space-region *k* there is a white raven'. In his formal requirement for test statements Popper does not pay attention to this.

The simplest solution to this problem is to change Popper's formal requirement for test statements. There is no reason why not all types of singular statements about observable events should be test statements. According to this modification not only the statement 'In the space-time-region *k* there is a planet' is a test statement, but also 'In the space-time-region *k* there is no planet', and even 'The planet Venus is in the space-time-region *k*' and 'The planet Venus is not in the space-time-region *k*'.

Popper introduces his formal requirement for test statements in order to guarantee that universal hypotheses are falsifiable. A universal hypothesis having the logical form of a general statement can be refuted by a test statement having the logical form of a singular existential statement. However, in order to falsify a universal hypothesis, it is not necessary to require that test statements should have the logical form of singular existential statements. If we modify Popper's formal requirement for test statements as suggested above, we can also falsify universal hypotheses with test statements. Furthermore, test statements are used not only in order to falsify isolated universal hypotheses, but also in order to falsify different types of theoretical systems and in order to falsify other test statements. With the help of metalogical considerations (in chap. 2, C, above and in the appendix below) it can be shown that all these types of falsifications are valid. Hence we can modify Popper's formal requirement for test statements in the way suggested here.

The simplest way to lay down the logical form of test statements is to consider their role as initial conditions and prognoses in scientific explanations and predictions. The most important function in science of singular statements about specific space-time-regions is their functions as initial conditions and as prognoses in scientific explanations. Potential falsifiers of theoretical systems are negations of prognoses, or negations of prognoses together with initial conditions (see chap. 2, C above). The function of test statements as initial conditions and prognoses is primary. Their function as potential falsifiers can be derived from this primary function.

The suggested modification of Popper's formal requirement for test statements makes it possible to falsify different types of theoretical systems and to falsify other test statements. Therefore it fulfils Popper's intentions. According to the suggested modification, test statements can have different logical forms. Which form of test statement we choose in a specific test situation depends on which type of test statements is unproblematic in that situation. In many cases test statements about specific individuals are unproblematic, as for example test statements about the position of the planet Venus. In such cases it is not necessary to use the logically weaker existential statements as test statements. (From the test statement 'The planet Venus is in the space-time-region *k*' follows the logically weaker existential statements 'There is a planet in the space-time-region *k*'.)

3. *Fallibility of Test Statements*

In addition to the formal requirement for test statements Popper lays down a material requirement: The events described in test statements should be observable, that is, test statements should be intersubjectively testable by observations (Popper 1959, § 28, 102) Popper does not require that test statements should be verifiable. In order to prevent such a misunderstanding, Popper speaks of 'observable' in the sense of 'intersubjectively testable'. Positivist philosophers of science require that test statements should be verifiable by observation (Andersson 1984, 50–52). Popper rejects this requirement because all descriptive statements go far beyond the immediate experience ("transcendence inherent in any description"): "Every description uses *universal* names (or symbols, or ideas); every statement has the character of a theory, of a hypothesis" (Popper 1959, § 25, 94–95). This is also true for the singular statements used as test statements. In order to describe observable events, they have to use universals, like 'planet', 'swan', or 'white'. Also the most simple test statements have to use universals and hence cannot be reduced to immediate experience. This is true even for "'phenomenal' language permitting statements like 'now here red'." It "would be impregnated with theories about time, space, and colour" (Popper 1959, ch. 5, addendum, 1972, 111). The positivist wish for an infallible empirical basis of science cannot be fulfilled. Not only test statements, but even experience itself is impregnated with theory. Without theoretical presuppositions neither knowledge nor experience is possible (cf. Popper 1979, 52).

Following Tarski, Popper maintains that there cannot be any general criterion of truth (Popper 1966, 2: 373 (Addendum 1 (1961)). Hence there cannot be any general criterion for the truth of test statements. Neither experience nor consensus, nor empirical corroboration is such a criterion.

Popper mentions a further argument for the fallibility of test statements: experience cannot prove the truth of test statements (or of any other statements). Statements can be proved only by other statements: "Experiences can *motivate a decision*, and hence cannot be *justified* by them—no more than by thumping the table" (Popper 1959, § 29, 105).

One of Popper's major concerns in *The Logic of Scientific Discovery* is to show that test statements are fallible. On this point there is agreement between Popper on the one hand and Kuhn and Feyerabend on the other. Lakatos's description of "dogmatic falsificationism" does not describe Popper's falsificationism because dogmatic falsificationism rests on the assumption that test statements are infallible.

The fallibility of test statements leads to a methodological problem. If experience cannot prove the truth of test statements and if there is no general criterion of truth for these statements, which test statements can be used in order to test theories? It is not enough to maintain, as for example Neurath did, that test statements are fallible, you also have to give rules showing how a critical discussion of test statements is possible:

> Neurath fails to give any such rules and thus unwittingly throws empiricism overboard. For without such rules, empirical statements are no longer distinguished from any other sort of statements. ... Neurath avoids one form of dogmatism [the dogmatism of infallible test statements], yet he paves the way for any arbitrary system to set itself up as 'empirical science' (Popper 1959, § 26, 97).

According to Popper, Neurath's view that test statements are fallible is a notable advance. But it leads nowhere if it is not followed up by another step solving the methodological problem of how to choose among the fallible test statements. Thus Popper criticises Neurath for not having solved the methodological problem of fallible test statements. It is ironical that Kuhn later criticises Popper in the same way as Popper earlier criticised Neurath (see chap. 3, A above).

The fallibility of test statements leads Neurath and later Kuhn and Feyerabend to relativism. Does Popper avoid this consequence? Has he shown how a critical discussion of fallible test statements is possible? Popper assumes "a realist and objectivist point of view" and tries "to replace *perceptions* as 'basis' by *critical testing* (Popper 1959, addendum 1972, 111). Does Popper's attempt succeed?

4. *Reproducibility of Test Statements*

Is it possible to replace perceptions as 'basis' by critical testing? Do not tests depend on perceptions? Popper tries to replace the idea that test statements can be verified by the idea that test statements can be critically tested. It is an expression of psychologism to think that test statements can be verified by perceptions (Popper 1959, §§ 2, 8, 27) Perceptions do not guarantee the truth of test statements. Nevertheless, Popper regards test statements as objective in the sense that they can be inter-subjectively tested:

> Only when certain events recur in accordance with rules or regularities, as is the case with repeatable experiments, can our observations be tested—in principle—by anyone. ... Indeed the scientifically significant *physical effect* may be defined as that which can be regularly reproduced by anyone who carries out the appropriate experiment in the way prescribed. No serious physicist would offer for publication, as a scientific discovery, any such 'occult effect', as I propose to call it—one for whose reproduction he could give no instructions (Popper 1959, § 8, 45).

Test statements should describe reproducible effects, effects that in principle can be repeated by any scientist. Test statements describing such effects are reproducible and inter-subjectively testable. The reproducibility of test statements requires that from one test statement other test statements can be deduced. Therefore "there can be no statements in science which cannot be tested, and therefore none which cannot in principle be refuted, by falsifying some of the conclusions which can be deduced from them" (Popper 1959, §8, 47).

Popper discusses the following example of a reproducible effect: A scientist observes a well-known chemical reaction in a test-tube and sees, as expected, a brightly green liquid. He observes how the colour for a short moment changes from brightly green to violet. After that the liquid once again has its well-known green colour (Popper 1979, 66–67). The following test statements are discussed:
(1) In the time-space-region *k* there is brightly green liquid. (*Pk*)
(2) In the time-space-region *l* there is violet liquid. (*Ql*)
(3) In the time-space-region *m* there is brightly green liquid. (*Pm*)

The test statements (1) and (3) describe the same event *P* (brightly green liquid) and are therefore called *homotypic* test statements by Popper. On the other hand the test statements (1) and (2) describe the different events *P* and *Q* (brightly green and violet liquid). These test statements can be called heterotypic.

An event is reproducible if the observation of the event can be repeated regularly: "Any empirical scientific statement can be presented (by describing experimental arrangements, etc.) in such a way that anyone who has learned the relevant technique can test it" (Popper 1979, § 27, 99).

Assume that we want to test the hypotheses 'All threads of structure *S* loaded with more than 2 *lbs.* break' (cf. chap. 2, A and B, above). From this hypothesis follows the test hypothesis 'All threads of structure *S* loaded with 4 *lbs.* break'. From this test hypothesis we can derive a number of conditional prognoses, as for example: 'If there is a thread of structure *S* in the time-space-region *k* and it is loaded with 4 *lbs.*, then it breaks.' For other time-space-regions similar predictions can be derived.

If empirical tests in the different time-space-regions show that a thread of structure *S* loaded with 4 *lbs.* regularly breaks, then the test effect is reproducible and the corresponding test hypothesis is corroborated. The corroboration of the tested hypothesis is reproducible.

If a thread of structure *S* is loaded with 4 *lbs.* and does not break, then the tested hypothesis is falsified from a logical point of view. But Popper does not regard this as a sufficient condition for the falsification of a theoretical system. He regards a theory as falsified only when a reproducible falsifying effect is found, that is when a low-level test hypothesis describing this effect is corroborated. Popper calls such test hypotheses falsifying hypotheses (Popper 1959, § 22) Such a falsifying hypothesis is for example 'No thread of structure *S* breaks, if it is loaded with 4 *lbs.*'. If this hypothesis is corroborated, then the falsifying effect is reproducible and the tested hypothesis falsified according to Popper's requirement that falsifications should be reproducible. It is not always easy to find a reproducible falsifying effect. It is for example possible that not all threads of structure *S* break if loaded with 4 *lbs.*, but only a subset *S´* of them. Then this falsifying subset *S´* has to be discovered and the corresponding test hypothesis 'No thread of structure *S´* breaks if it is loaded with 4 *lbs.*' has to be formulated and tested.

It is not necessary to put forward a corroborating test hypothesis in order to corroborate a theory. Such test hypotheses do not have to be found or discovered, but can be derived directly from the tested hypothesis. For example, if we test the hypothesis 'All threads of structure *S* loaded with more than 2 *lbs.* break', then it has the test hypothesis 'All threads of structure *S* loaded with 4 *lbs.* break' as a consequence. Test hypotheses used to falsify (falsifying hypotheses) have to be discovered. If we once observe that a thread of structure *S* does not break when loaded with 4 *lbs.*, it is not certain that the thread will break again if the experiment is repeated. If it does not break, the test hypothesis is falsified and the test effect is not reproducible (Popper 1959, § 8, 45).

Falsifications and corroborations should be reproducible. The reproducibility of test effects can be tested if test statements of the same type (homotypic test statements) can be derived from the tested hypothesis or from a falsifying hypothesis. In science we are not interested in stray test statements which are logically disconnected, but in test statements connected with reproducible effects and used to corroborate or falsify scientific theories (Popper 1959, § 30, 106).

Objective knowledge is not infallible knowledge, but inter-subjectively testable knowledge. Since test statements describe special space-time-regions and thus are singular statements, the requirement that test statements should be inter-subjectively testable "can of course only refer to observers who are suitably placed in space and time," (cf. Popper 1959, § 28, 102–3; Popper 1979, 67).

However, the requirement that test effects should be reproducible goes further and refers not only to observers in special space-time-regions. If the test effect is reproducible, then anybody should in principle be able to reproduce the test. The corresponding test hypothesis need not be strictly universal, but can for example refer to a special space-region. Such a test hypothesis is 'There is a family of white ravens in the zoo at New York', which can be used in order to refute the hypothesis 'all ravens are black' (Popper 1959, § 22, fn. 1) and which makes the test effect 'white raven' reproducible.

Irreproducible (occult) effects suggest problems which can stimulate research, as for example reports on sea-serpents. "In such cases science does not say that the reports are unfounded... or that the reports of observed sea-serpents are lies. Instead, it suspends judgement" (Popper 1959, § 27, 100). It is not sufficient to repeat an observation some times in order to make it reproducible, but there must be a test hypothesis telling us how to reproduce the effect regularly.

For example, there are many observational reports on the monster in Loch Ness. The observers are described as intelligent and reliable. However, in spite of the efforts of many research groups, nobody has succeeded in making the observations of the monster reproducible. Nobody has for example succeeded in catching an exemplar of the monster or in telling how it can be observed regularly. From a scientific point of view, the monster in Loch Ness is an open problem.

An interesting counterhypothesis is that in reality the monster is a trunk of the Scotch pine (*Pinus sylvestris*). These trees grow along the shore of Loch Ness. If they fall into the sea, they sink to the bottom of the deep sea. There the high pressure will press resin and phenols from the trunk to form an outside casing. Due to decay processes gases are formed and bubble out to form blisters in the casing. These blisters become buoyancy tanks, and after a time the trunk will rise to the surface of the sea with high speed. But as the trunk comes nearer the surface, the outside pressure decreases, and the gas blisters become bigger and eventually explode. Therefore the trunk will sink back to the bottom of the sea, after having reached the surface with high speed and with a hiss of escaping gas. This hypothesis is intersubjectively testable:

> The course of future research is clear: lower a grab and bring a sample log to the surface. This should not be difficult for there are hundreds down there. ... Then, perhaps, a proper research project could be mounted by sinking a weighted log to simulate waterlogging, the log being fitted with transducers, to record pressure and the rest (Craig 1982, 356–57).

If tests showed that the Scotch pine really behaves in this way, this could solve the open problem and demystify the monster in Loch Ness.

5. *Deduction of Heterotypic Test Statements*

Every empirical test has to stop at some test statements, which are tentatively maintained in the critical discussion. According to Popper some test statements have to be "accepted" (Popper 1959,§ 29, 69). In *The Logic of Scientific Discovery* Popper does say that test statements should be verified, as he (1979, 125, 261) did in *Die beiden Grundprobleme der Erkenntnistheorie*. However, in this work it is also clear that Popper regards test statements as fallible (Popper 1979, 261, fn. *1, 126–28). The fallibility of test statements has the consequence that they can only be accepted tentatively and temporarily and are always open to criticism.

Bartley finds it superfluous to say that test statements are accepted and even thinks that this way of speaking is a justificationist residue (Bartley 1982, 166). However, if we keep in mind that test statements are not accepted as true, but that they are always fallible and criticisable, then no justificationist problems should arise. Bartley thinks that Popper has "an unfortunate tendency to demand convention or irrational decision whenever some point is reached which cannot be justified." Bartley proposes "simply to eliminate as superfluous Popper's requirements relating to decisions, agreements, conventions, acceptance, and justification" (Bartley 1984, 215). Test statements need not be verified or accepted as true, but "are intended to be hypothetical, and criticisable and revisable, just like everything else in the system" (Bartley 1984, 215). The problem of how to justify test statements is replaced by the problem of how to criticise them (cf. Albert 1991, § 6). If test statements withstand criticism, they can be "accepted" tentatively.

If some tested prognoses are problematic, usually other and less problematic prognoses are derived from the theory; if initial conditions for tests are problematic, often other and unproblematic conditions can be chosen. Therefore, there are relatively few cases in which we have to discuss a specific type of test statements (Popper 1979, 125–26).

Nevertheless, it is possible to criticise specific test statements. Popper discusses the following statement: 'This powder is red precipitate'. If this statement is problematic, other and less problematic statements can be derived from it with the help of auxiliary hypotheses, as for example the following conditional prognosis: 'If red precipitate is strongly heated in a test tube, then oxygen is formed and metallic mercury collects on the walls of the test tube'. If the initial condition is satisfied, that is, if the test tube is strongly heated, then new prognoses can be tested: 'There is oxygen in this test tube' and 'there is mercury on the walls of this test tube'.

If these test statements should also be problematic, the derivation of test statements can be continued. For example, we can test whether there is oxygen in the test tube by appropriate tests. According to an auxiliary hypothesis red-hot iron cuttings flame up, if introduced into oxygen. Therefore we can derive the prognosis that red-hot iron cuttings will flame up, if introduced into the test tube (Popper 1979,125–26).

This example shows how new types of test statements (heterotypic test statements) can be derived with the help of auxiliary hypotheses. The derivation has the following logical form:

In the space-time-region *k* there is a *P*. (*Pk*)
Auxiliary hypotheses. (*AH*)
―――――――――――――――――――――――――
In the space-time-region *l* there is a *Q*. (*Ql*)

or

(1) $Pk, AH \vdash Ql$

From test statements about red precipitate we can derive test statements about mercury, oxygen, and flaming iron cuttings. In actual testing we prefer unproblematic test statements, which can easily be tested inter-subjectively.

In a similar way, we can also deduce homotypic test statements:

(2) $Pk, AH \vdash Pl$

Tests statements are often tested with homotypic test statements without much reflection simply by repeating observations or experiments. If a scientist believing that all swans are white sees a black swan, his first reaction is probably to look once again to test whether the swan really is black. "We do not take even our own observations quite seriously, or accept them as scientific observations, until we have repeated and tested them" (Popper 1959, § 8, 45). After such tests a test hypothesis has to be proposed in order to make the observations of black swans reproducible, for example a hypothesis telling the scientists where black swans can regularly be observed. In this way it becomes inter-subjectively testable that there are black swans.

Test statements can be tested by the derivation of further test statements. Therefore there are no absolutely final test statements or 'basic statements'. From a logical point of view, we are never forced to stop at a particular type of test statement. In actual research we use such test statements that can easily be tested inter-subjectively. Often auxiliary hypotheses are implicitly presupposed, and the test statements are theory impregnated and far away from "pure" observations. If the test statements presuppose problematic background knowledge, then other and less problematic test statements have to be derived. Any test statement can in its turn be subjected to tests (Popper 1959, § 29, 104).

In actual research scientists are using test statements that are unproblematic for them, but not for the layman. Pierre Duhem discusses a layman seeing only the movements of an iron bar carrying a mirror, when a scientist maintains the he is measuring the electrical resistance of a coil. If the layman asks the scientist how this is possible, the scientist would answer that this requires "some very long explanations" and would recommend the layman to take a course in electricity (Duhem 1974, 144).

A test statement can be tested with other (homotypic or heterotypic) test statements if we assume auxiliary hypotheses. In this way isolated test statements can be discussed and subjected to further tests. However, test statements are seldom discussed in isolation, but in relation to some theory to be tested with them. In such cases we simply derive different and unproblematic prognoses from the tested theory, if some type of prognosis should be problematic. Occasionally we have to introduce new auxiliary hypotheses in order to be able to derive unproblematic prognoses. If initial conditions should be problematic, the simplest way is to choose different and unproblematic initial conditions in order to derive testable consequences. Therefore, deductive tests of isolated test statements are rather unusual. However, for a methodological discussion of the problem of test statements, it is important that such tests are always possible.

Which test statements can be used in the critical discussion of theories? We should use such test statements that are unproblematic and easy to test intersubjectively.

> [W]e are stopping at statements about whose acceptance or rejection the various investigators are likely to reach agreement. And if they do not agree, they will simply continue with the tests, or else start them all over again. If this too leads to no result, then we might say that the statements in question were not inter-subjectively testable, or that we were not, after all, dealing with observable events (Popper 1959, § 29, 104).

B. KUHN ON TEST STATEMENTS

Kuhn (1970a, 15) thinks that Popper "is entirely silent" on the problem of test statements. As was shown above, this is not the case. Popper not only maintains that test statements are fallible and theory-impregnated, but also provides an interesting solution to the problem of how to discuss fallible test statements critically. Therefore, Kuhn's (1970a, 14) idea that Popper "legitimately can be treated" as a naive falsificationist believing in infallible test statements and conclusive disproofs of theories is mistaken. The problem is not, as Kuhn thinks, that Popper does not say anything on how to discuss fallible test statements, but whether Popper's criticist solution to the problem of empirical testing is acceptable.

However, first I will discuss a logical problem. If test statements are fallible, then falsifications must also be fallible and inconclusive. But "what is falsification if it is not conclusive disproof" (Kuhn 1970a, 15)?

1. *Falsification with Fallible Test Statements*

If test statements are fallible, then falsifications cannot be conclusive disproof of hypotheses or absolutely certain justification of their falsity. It is a justificationist misunderstanding to regard falsifications as infallible. Such a position cannot be defended from a fallibilist point of view because not only general hypotheses, but also singular test statements and hence also falsifications are fallible. We cannot assert with absolute certainty that a test statement is true or a general hypothesis is false.

But what is a falsification, if not conclusive disproof? It is a conditional falsification asserting that *if* some test statements (potential falsifiers) are true, *then* a general theory is false. Thus every falsification is inconclusive in the sense that it is conditional and depends on the truth of some test statements (or potential falsifiers). It can never be asserted with absolute certainty that test statements, potential falsifiers, or falsifying premises are true. Therefore falsifications are as fallible as the test statements used as premises in the falsification.

Falsifications are not conclusive disproof, but conditional disproof. It is logically true that if a potential falsifier of a theory is true, then the theory is false (Andersson 1982). For example, if the potential falsifier 'There is a non-white swan in the space-time-region *k*' is true, then it follows conclusively and with logical necessity that the hypothesis 'All swans are white' is false. However, the test statement that there is a non-white swan is fallible, even if it is accompanied by a falsifying hypothesis telling us how to reproduce the falsification. Therefore the statement 'Not all swans are white' is also fallible. Even if falsifications are seldom revised (Popper 1959, § 82, 268), such revisions are in principle always possible, and falsifications are never conclusive or final.

2. *Incommensurability*

Kuhn maintains, like Popper, that test statements are revisable. According to Popper this is a consequence of the fallibility of test statements. Kuhn motivates the revisability of test statements in another way using psychological arguments. Experiments in gestalt psychology show that different gestalts can be seen when the same picture is observed. Wittgenstein (1967, part 2, § xi) discusses the following example:

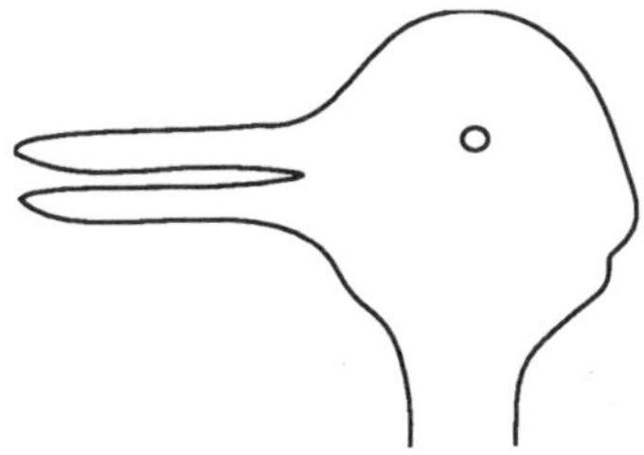

Figure 1

You can see a duck or a rabbit when observing this figure. It is possible to maintain the different test statements 'At *k* there is a duck' and 'At *k* there is a rabbit'.

Kuhn argues that in this way you can also see different gestalts when observing or experimenting in science. He thinks that the paradigm determines which gestalt you actually see (cf. chap. 3, A, above). For example, if you use a duck-paradigm, you expect to see ducks and do not only interpret observations as ducks, but immediately see ducks without any conscious interpretation. In the same way scientists immediately see phlogiston in a chemical experiment, if they use a phlogiston-paradigm, or oxygen if they use an oxygen-paradigm. The paradigm determines what test statements are actually maintained, and independent tests of scientific theories are impossible. Hence scientific theories are incommensurable. In order to discuss this problem, I will analyse some of Kuhn's historical examples of incommensurable theories.

2.1. *The Discovery of Uranus*

Kuhn discusses Sir William Herschel's discovery of Uranus in 1781 as an example of the change of the "world view" accompanying scientific revolutions (Kuhn 1970b, 115–17). Uranus was observed before 1781, but before Herschel's discovery astronomers thought they observed a star. Kuhn compares the discovery of Uranus to a gestalt switch: first the astronomers saw a star and then they saw a planet when observing Uranus. Before 1781 the astronomers maintained the test statement 'At *k* there is a star', and afterwards 'At *l* there is a planet'. Thus they maintained heterotypic test statements before and after Herschel's discovery.

Kuhn explains this "shift of scientific vision" as the result of a minor change of paradigm accompanying Herschel's discovery. However, Kuhn stresses that the paradigm shift was not the cause of the discovery, but rather a consequence of it. The cause of the discovery was some "anomalies" seen by Herschel. With a much improved telescope Herschel was able to see that Uranus looks like a disk rather than a point and that

Uranus moves among the stars, observations that did not fit to the star Gestalt, but were unexpected and anomalous. These "anomalies" were the reasons why Herschel thought that "something was awry" (Kuhn 1970b, 115). What was awry? Herschel writes:

> In examining the small stars in the neighbourhood of H *Geminorum* I perceived one that appeared visibly larger than the rest; being struck with its uncommon appearance I compared it to H *Geminorum* and the small stars in the quartile between *Auriga* and *Gemini*, and finding it so much larger than either of them, I suspected it to be a comet. (Berry 1961, 325, quoting Herschel).

In order to understand this it is useful to regard the following prediction:

In the space-time-region *k* there is a star. (*Pl*)
The object in the space-time-region *k* is observed with a telescope. (*Qk*)
All stars are so far away that they are punctiform when observed with a telescope. (*AH*)

The object in the space-time-region *k* is punctiform. (*Rk*)

or

(3) $Pk, Qk, AH \vdash Rk$

With his much improved telescope Herschel was able to observe that Uranus appears like a disk. He was able to maintain the test statement that Uranus is not punctiform ($\neg Rk$). Since the conclusion of the derivation above is false, at least one of the premises is false. If we assume that the auxiliary hypothesis (*AH*) is true and that Uranus was observed with a telescope (*PK*), then the remaining premise must be false, that is, Uranus is not a star ($\neg Pk$). Thus the following falsification is valid (theorem 3d in the appendix below, $r = 1, n = 2, k = 1$):

(4) $\neg Rk, Qk, AH \vdash \neg Pk$

In this falsification an auxiliary hypothesis is assumed. To use general hypotheses among falsifying premises to refute an initial condition is unproblematic from a logical point of view. If (3) is valid, then (4) is also valid. From a methodological point of view this more general use of falsifications is interesting and useful. You cannot only falsify general hypotheses, but also initial conditions. Whether you test a hypothesis or an initial condition depends on what you regard as problematic. In this case the auxiliary hypothesis is regarded as unproblematic and one initial condition as problematic.

The falsification (4) shows how a test statement can be criticised with other test statements. From a problematic test statement another unproblematic and heterotypic test statement is derived: from the test statement that there is a star at *k*, we can derive the heterotypic test statement that there is a punctiform object at *k*. We can also, conversely, derive that if there is a non-punctiform (disklike) object at *k*, then there is no star at *k*.

Kuhn calls the disklike appearance of Uranus an anomaly. It is interesting that the corresponding test statement is a falsifying heterotypic test statement, as (4) shows.

Although it is possible to understand the discussion about Uranus as a discussion of singular test statements, it is more realistic to understand it as a discussion of the (low-level) hypothesis 'Uranus is a star'. From this hypothesis you can derive the following test statement:

> In the space-time-region *k* Uranus is observed with a telescope. (*Qk*)
> Uranus is a star. (*H*)
> All stars are so far away that they are punctiform when observed with a telescope. (*AH*)
> ---
> In the space-time-region *k* Uranus is punctiform. (*Rk*)

This prediction allows us to test the hypothesis that Uranus is a star. If observation in a telescope shows that Uranus looks like a disk and if the auxiliary hypothesis is true, then we can derive that Uranus is not a star, that is, the tested hypothesis is refuted (according to theorem 3d, $r=2$, $n=m=1$, $k=0$, in the appendix below):

> In the space-time-region *k* Uranus is observed with a telescope. (*Qk*)
> All stars are so far away that they are punctiform when observed with a telescope. (*AH*)
> In the space-time-region *k* Uranus is not punctiform. ($\neg Rk$)
> ---
> Uranus is a not star. ($\neg H$)

This falsification is reproducible. The falsifying hypothesis 'When Uranus is observed with a powerful telescope, then Uranus is not punctiform' is inter-subjectively testable and corroborated.

Thus the first step in the discovery of the planet Uranus is a falsification of an earlier hypothesis, the hypothesis (and the corresponding test statements) that Uranus is a star. What is Uranus if not a star? First Herschel thought that he had discovered a comet. When comets are far away from the sun, they do not develop any tail and can be disklike, like a planet. Since new comets had earlier been discovered, Herschel assumed that the disklike object he observed was a comet. Herschel withdrew this hypothesis about the nature of Uranus only when some astronomers calculated the orbit of Uranus and found that it did not move like a comet, but in an almost circular orbit around the sun, like a planet (Berry 1961, 325–26; Wolf 1890–92, 2: 477).

It is also possible to refute the hypothesis that Uranus is a star without a powerful telescope. A further anomaly to the star hypothesis is the (daily) motion of Uranus among the stars, which can also be observed with a small telescope. Herschel observed the motion of Uranus in the first evening of observation (Wolf 1890–92, 2: 477).

Kuhn calls the disklike appearance and the daily motion of Uranus anomalies. They can be described by test statements falsifying the hypothesis that Uranus is a star. In this way the gestalt switch from star to planet can be explained as the result of a falsification of the star-hypothesis. What Kuhn calls a conversion to the planet-hypothesis must not be regarded as irrational, but as the result of a critical discussion leading to the refutation of a low-level hypothesis, or test statement. Herschel's discovery of Uranus shows that a critical discussion of low-level hypotheses and test statements is possible. To maintain that paradigms or theoretical presuppositions alone determine which test statements are accepted is exaggerated and an expression of the myth of the framework. Kuhn's example rather shows that gestalt switches are triggered by anomalies, that is, by falsifying test statements.

2.2. *The Copernican Revolution*

Kuhn's paradigmatic case of a scientific revolution is the Copernican revolution. Here it is not easy to find any gestalt switches or incommensurable observations. The observations of the positions of the planets did not change. In this respect the Ptolemaic and Copernican theories were commensurable.

Kuhn, however, maintains that the observations of the moon changed with the Copernican revolution. The Ptolemaic astronomers regarded the moon as a planet and saw this gestalt when observing the moon, while the Copernican astronomers regarded the moon as a satellite of Earth and saw another gestalt (Kuhn 1970b, 115). (According to the Ptolemaic theory the moon is a planet and moves around Earth like the other planets; according to the Copernican theory the moon is the only satellite of Earth.)

This example is not convincing. You cannot see directly that the moon is a planet or a satellite, only that the moon moves among the stars. This movement can be explained with different hypotheses about the movement of the moon. Even if the Copernican and Ptolemaic theories differ in detail, both assume the moon moves around Earth and hence that the moon is a satellite of Earth. Thus there cannot be any gestalt switch when the Copernican theory was introduced (cf. Andersson 1981b, 257–59, 264–69).

2.3. *The Chemical Revolution*

Kuhn maintains that theories separated by scientific revolutions are incommensurable. But in all the examples of such theories discussed by him it is possible to find test statements common to both theories and unproblematic, that is, to find perfectly commensurable test statements. For example, let us discuss the scientific revolution leading from the phlogiston theory to the oxygen theory. According to Kuhn, Lavoisier, who introduced the oxygen theory, saw nature in a new way, at the same time as Priestly, who defended the old phlogiston theory, saw it in the old way: "Lavoisier… saw oxygen where Priestley had seen dephlogisticated air [air without phlogiston]" (Kuhn 1970b, 118).

Let us regard the two test statements 'In the space-time-region *k* there is oxygen' and 'In the space-time-region *k* there is dephlogisticated air'. Did Lavoisier and Priestley immediately see oxygen or dephlogisticated air or, at least, the corresponding gestalts?

In the beginning of their experiments Lavoisier and Priestley saw neither dephlogisticated air nor oxygen, but a colourless gas formed when "red precipitate" (a red oxide of mercury) was heated. First both thought that a colourless gas, at that time known as fixed air, today called carbon dioxide, was formed. This conjecture was immediately refuted because the new gas was not easily soluble in water, as carbon dioxide is. In the beginning only a colourless gas was observed, which was not so easily soluble in water as carbon dioxide.

Let us assume for the sake of discussion that Priestley and Lavoisier first "saw" carbon dioxide (fixed air) and maintained the test statement: "In the space-time-region *k* there is carbon dioxide (fixed air)." Assuming that carbon dioxide is easily soluble in water (auxiliary hypothesis, part of the background knowledge), we can easily test this test statement (or low-level hypothesis) by deriving a new kind of (heterotypic) test statement:

In the space-time-region *k* there is carbon dioxide (fixed air). (Pk)
Carbon dioxide (fixed air) is easily soluble in water. (AH)
———
In the space-time-region *k* there is a gas easily soluble in water. (Qk)

If the auxiliary hypothesis is unproblematic and the gas examined is not easily soluble in water, we can infer that it is not carbon dioxide:

In the space-time-region *k* there is no gas easily soluble in water. ($\neg Qk$)
Carbon dioxide (fixed air) is easily soluble in water. (AH)
———
In the space-time-region *k* there is no carbon dioxide (fixed air). ($\neg Pk$)

The test statements about the solubility of the new gas were unproblematic and accepted both by Lavoisier and Priestley. Both thought that the new gas could not be carbon dioxide (fixed air) (Conant 1957, 74–88).

In this way several hypotheses about the nature of the new gas were refuted. Priestley repeats how surprised he was when finding that the new gas had new and unexpected properties (Conant 1957, 99). An experiment performed by Priestley was to investigate how long a mouse could live in the gas. He observed that it could survive longer than in the same amount of common air. Further experiments showed that the new gas had other properties different from common air. Only after these experiments did Priestley conjecture that he had discovered a new gas. As a follower of the phlogiston theory he assumed that dephlogisticated air was formed in the experiment (he did not see it). This was the decisive step in discovering what is today called oxygen (Conant 1957, 100–102).

Even before Priestley's discovery Lavoisier had questioned the phlogiston theory, which states that combustion is accompanied by the delivery of phlogiston from the combusted substance to the air. Hence combustion should be accompanied by loss of weight. But Lavoisier saw that in some experiments with sulphur and phosphorus combustion is accompanied by an increase in weight. Therefore he assumed that something is taken from the air in combustion. First he thought that this substance was common air in an unusually pure condition. Also Priestley had earlier thought that common air is formed when red precipitate is heated. When Lavoisier heard about Priestley's further experiment he assumed that the new gas Priestley called dephlogisticated air was necessary for combustion. Lavoisier called this gas oxygen (Conant 1957, 78–88).

Priestley and Lavoisier saw neither dephlogisticated air nor oxygen, but assumed hypothetically that these gases are formed in special experiments. In order to be able to discover oxygen (dephlogisticated air) a long process of learning was necessary. During this process several hypotheses about the new gas were refuted. The test statements used in these refutations were unproblematic and accepted both by Lavoisier and Priestley, as for example the test statement about the solubility of the new gas in water. Thus Lavoisier and Priestley *saw* the same things and used the same test statements, but *explained* them differently, Lavoisier with the oxygen theory and Priestley with the phlogiston theory.

The gestalt-psychological model of experience brings Kuhn to regard explaining hypotheses as immediate observations. Lavoisier and Priestley explained chemical experiments in different ways. But they neither accepted different test statements nor lived in different worlds, as Kuhn suggests.

In this way all Kuhn's historical case studies of incommensurable theories can be analysed. Kuhn treats hypothetical explanations as immediate experiences. With a fallibilist and criticist view of test statements we can understand the history of science better than with Kuhn's model of experience influenced by gestalt-psychology.

2.4. *Gestalt Psychology and Criticism*

Kuhn maintains that there is no satisfactory alternative to the traditional epistemological paradigm, but that nevertheless all attempts to introduce a neutral observational language seem deemed to fail (Kuhn 1970b, 126). The traditional paradigm assuming infallible test statements is in a crisis, but a new and satisfactory epistemological paradigm is not yet available.

Put in another way, the traditional view of infallible test statements is highly problematic, but a satisfactory alternative is unavailable.

Kuhn (1970b, postscript) has weakened the gestalt-psychological model of experience and the thesis of incommensurability. He no longer maintains that electrons, electric currents, and so on, can be seen directly, but only tracks or else bubbles of vapour in a cloud chamber or the needle of an ammeter or galvanometer (Kuhn 1970b, 196–97):

> The metaphor that transfers 'seeing' to contexts like these is scarcely a sufficient basis for such claims. In the long run it will need to be eliminated in favour of a more literal mode of discourse. ... It ... requires contemplation, analysis, and interpretation (or else the intervention of external authority) before conclusions can be reached about electrons or currents.

Here Kuhn falls back on the position that electrons and currents are interpretations of experience ("seeing as"). This retreat does not lead to any new solution to the problem of test statements. Kuhn informs us that he is working with a computer program, which begins to suggest such a solution, but he does not give any details (Kuhn 1970b, 197). Thus the problem is still unsolved.

Kuhn uses historical cases in order to show that experience is theory-impregnated and that competing theories are incommensurable. Popper also thinks that experience is theory-impregnated, but he does not draw the radical conclusion that competing theories are incommensurable. On the contrary, Popper rejects this idea as an expression of "the myth of the framework" (Popper 1970b, 55–57; Popper 1976, 39–40). In order to be able to discuss the problem of incommensurability, I will again discuss the discovery of Uranus.

According to a strong interpretation of the thesis of incommensurability, you have to maintain that before Herschel's discovery there was a star where later there was a planet. Although Kuhn later weakens this position, he still maintains that a conversion is central to every scientific revolution (Kuhn 1970b, 204). Was a critical discussion, as Popper maintains, or a conversion, as Kuhn maintains, the basis for the small scientific revolution leading to the discovery that Uranus is a planet?

The test statements about Uranus discussed above were commensurable. You did not have to convert in order to be able to accept them. What Kuhn calls anomalies, the shape and movement of Uranus, are new unproblematic and inter-subjectively testable effects, which can be critically discussed. They make it possible to discuss the earlier suggested and problematic test statements or test hypotheses saying that Uranus is a star. Herschel's observation of the disklike appearance of Uranus falsifies the statement that Uranus is a star; his observations of the way Uranus moves falsify the statement that Uranus is a comet. These statements about Uranus were falsified with unproblematic test statements. No irrational conversion was necessary, only observation and deductive logic.

The gestalt-psychological model of experience is correct in so far as it points out that experience is theory-impregnated. But it is incorrect if it maintains that theories or paradigms are incommensurable. You do not have to treat perceptions of gestalts as final and uncriticisable experiences, but can treat them as criticisable hypotheses. There is no sudden break between hypothesis and experience.

According to Gregory perceptions can be regarded as hypotheses:

> It is possible ... that I am wrong that the dark blue patch, apparently over there, is the sea; or even that this which I see and touch is indeed my desk. I might possibly be dreaming or drugged. This may be unlikely but it is possible; so my perceptions are not *certain*. ... There seems to be no sudden break between *perceiving* an object and *guessing* an object. If all perceiving of objects requires some guessing, we may think of sensory stimulation as providing *data* for *hypotheses* concerning the state of the external world. The selected hypotheses, following this view, are perceptions (Gregory 1973, 61–62).

Campbell (1974, 419–37) thinks that "a nested hierarchy of selective-retention processes" is typical for all types of human knowledge, including perceptions. There also has to be some consistent selection process for perceptions. Kuhn and Feyerabend do not find any such processes and hence maintain the thesis of incommensurability. In the following I will argue that critical discussion of test statements provides such a consistent selection process.

Perceptions of gestalts are implicit hypotheses and can very well be criticised in a process involving the deduction of new and unproblematic consequences. What Kuhn calls 'anomalies' are details that do not fit to a gestalt and that hence can be used in a critical discussion of the perceived gestalt.

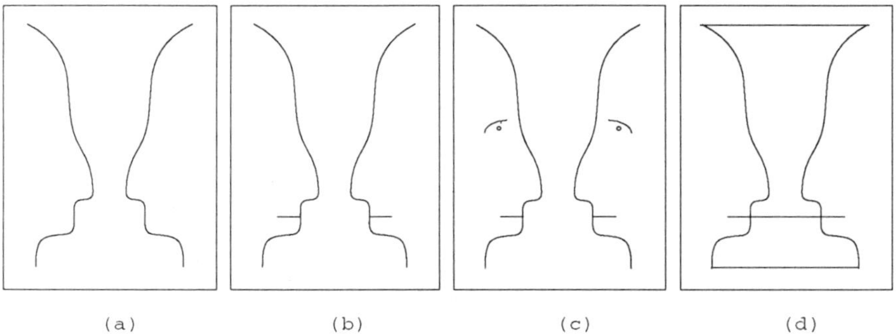

Figure 2

Figure 2a is ambiguous and can be regarded as a vase or as two faces. Two different gestalts can be seen. In a similar way Uranus could be seen in different ways: as a star, a comet, or a distant planet. Since three perceptions (gestalts) or hypotheses were possible, the corresponding test statements about an observed star, planet or comet were problematic. Further and unproblematic effects about the size and movement of Uranus had to be derived. In this way the problematic hypotheses and test statements about Uranus could be critically discussed.

The disklike appearance of Uranus refutes the hypothesis (and corresponding test statements) that Uranus is a star, but can be explained by the hypotheses that Uranus is a comet or a planet. In a similar way figure 2a above can be interpreted in two ways. Although figure 2b is more detailed, it is nevertheless ambiguous and can be interpreted in two ways. The new detail in figure 2b can be interpreted as part of a vase or as part of a face. However, figure 2c can only be interpreted as two faces. In a similar way, exact observations of the movements of Uranus only make one interpretation possible, the interpretation that Uranus is a planet.

The deduction of new types of test statements (heterotypic test statements) can be compared to the observation of new details in a picture. Perceptions of gestalts are hypotheses, even if they are immediate from a psychological point of view. Like other hypotheses, they can be tested by their deductive consequences. Therefore, a gestalt-psychological model of experience need not be incompatible with a criticist one. This is the case only when perceptions of gestalts are wrongly regarded as uncriticisable.

Often different hypotheses and perceptions of gestalts are possible. In science there are often different explanations and interpretations of experience. For example, in the beginning different test hypotheses about Uranus were possible.

If there is no agreement on test statements, then the test statements are problematic and the scientists must try to deduce new and unproblematic test statements. If scientists disagree on the nature of Uranus, whether it is a star or planet, they have to find new test statements on which they can agree, as for example on the size and movement of Uranus. On this new level of discussion intersubjective agreement is possible.

Critical rationalists do not seek agreement on test statements because they think that agreement is a criterion of truth, but because agreement shows which test statements are unproblematic. Such agreement does not rule out error, but only shows that we, for the time being, do not have any reasons to question the discussed statements.

Kuhn deals with some problematic test statements, which can be interpreted in different ways in his historical case studies. But, as was shown above, it is possible to derive simpler test statements on which intersubjective agreement can be reached. Observers could agree on the size and movement of Uranus, in spite of their different theoretical expectations or paradigms. When such unproblematic test statements are available, the problem of incommensurability does not arise. Hence, to solve the problem of incommensurability, we have to derive new and unproblematic test statements on which inter-subjective agreement can be reached. In all the examples of incommensurable theories discussed by Kuhn, such a derivation of unproblematic test statements is possible. Hence Kuhn's idea that many competing theories are incommensurable is an expression of the myth of the framework.

Unproblematic test statements are fallible and theory-impregnated. But they do not depend on problematic theories. The test statements about the size and movement of Uranus presuppose some rudimentary theories on form and movement of astronomical objects, but not the problematic hypotheses on the nature of Uranus. The test statements on the form and movement of Uranus were intersubjectively testable and generally accepted. In the same way the adherents of the Ptolemaic and Copernican theories could agree on the positions of the planets and the moon. These test statements were unproblematic and a common basis for discussion. In the discussion of the phlogiston and oxygen theories there were also many unproblematic and common test statements, such as, for example, that a colourless gas is formed when red precipitate is heated and that this gas is not easily soluble in water. Only the question of how these different unproblematic test statements should be explained was discussed.

C. LAKATOS ON TEST STATEMENTS

1. *Test Statements in Dogmatic Falsificationism*

Popper rejects one of the basic assumptions of dogmatic falsificationism explicitly, namely the assumption that experience can prove test statements. Is it true, as Lakatos maintains, that Popper was a dogmatic falsificationist before he published anything? After the publication of *Die beiden Grundprobleme der Erkenntnistheorie* (written before *The Logic of Scientific Discovery*, but published much later) it is easier to discuss this question. Although Popper in *Die beiden Grundprobleme* writes about verification and falsification of singular factual statements (Popper 1979, 260), he makes it clear that they are not verified or falsified by experience, but that scientists decide to accept them in a fallible way (Popper 1979, 126–27). This is the same position as in *The Logic of Scientific Discovery*. From the very beginning Popper criticises the positivist idea of an infallible empirical basis of science.

2. *Test Statements in Naive Methodological Falsificationism*

In naive methodological falsificationism scientists make test statements unfalsifiable by conventional decisions (cf. chap. 4, B, above). In this way Lakatos interprets what Popper writes about test statements in *The Logic of Scientific Discovery*. There Popper writes that test statements are accepted by conventional decisions, which from a logical point of view are free or arbitrary (Popper 1959, § 30, 108–9). Lakatos is not alone in interpreting these passages as a conventionalist solution of the problem of test statements. However, in these passages Popper wants to criticise the positivist idea that test statements can be verified by experience. He writes that the decision to accept a test statement is a free or arbitrary decision from a logical point of view. This only means that circumstances outside logic determine which test statements are accepted or rejected. Experience, for example, is such a circumstance. Popper does not mean that the decision to accept or reject a test statement is free or arbitrary from an empirical point of view, or that decisions can make test statements true or false, which would be absurd for a realist to maintain. Popper constantly stresses the inter-subjective testability of test statements. He tries to substitute empirical criticism for empirical proof. All test statements, even the accepted ones, remain fallible and criticisable. Later Popper has added: "I assume a realist and objectivist point of view: I try to replace *perception* as 'basis' by *critical testing*" (Popper 1959, addendum 1972, 111). This kind of critical realism is not a brand of conventionalism, but a new type of solution to the problem of test statements, a solution neither understood nor discussed by Lakatos.

Equally untenable is Lakatos's idea that test statements can be made unfalsifiable. No statement, not even a singular test statement, can be made unfalsifiable by a decision. A scientist can only decide not to test it. From a criticist point of view such a decision is dogmatic, even if it is called a conventionalist or methodological decision.

3. *Test Statements in Sophisticated Methodological Falsificationism*

In sophisticated methodological falsificationism there is an appeal procedure against the verdicts of scientists about test statements. Thus test statements are no longer unfalsifiable, as in naive methodological falsificationism, but criticisable.

How can test statements be criticised according to sophisticated methodological falsificationism? What does the appeal procedure look like? Lakatos rejects Popper's idea that test statements are criticised by their deductive consequences and thinks that they are criticised indirectly, with the help of interpretative theories (Lakatos 1970, 128).

Lakatos discusses the following example. According to Prout's theory all atoms are compounds of hydrogen atoms and thus all atomic weights of chemical elements must be expressible as whole numbers. This theory was falsified when Stas observed that the atomic weight of chlorine was not a whole number, but 35,5. Lakatos thinks that Prout's theory survived this refutation because Stas's refutation was criticised.

How can the test statement that the atomic weight of chlorine is 35,5 be criticised? Lakatos thinks that an interpretative theory saying that the sample of chlorine was chemically pure was criticised. Of course, if Prout's theory is about chemically pure chlorine and Stas experimented with chemically impure chlorine, we can appeal against Stas's refutation.

Unfortunately, Lakatos's example rests on a misunderstanding. Prout's theory did not survive because Stas's refutation was wrong or because he experimented with chemically impure substances, but because Prout's theory was later used to explain the atomic weight of isotopes. In the new form Prout's theory maintains that the atomic weights of homogenous isotopes are whole numbers. But for chemically pure substances Prout's theory is still false, as Stas correctly showed. Against this refutation there has not been any successful appeal: chemically pure chlorine found in nature still has the atomic weight 35,5. Thus there was no criticism of Stas's test statements or of his "interpretative theories," but of Prout's theory, which was later modified to a theory about isotopes, not about chemically pure substances (Aston 1938). The discussion of Prout's hypothesis only shows that falsified theories might be modified.

In the discussion of the oxygen and phlogiston theories chemically impure substances played a role, but not in the discussion of Prout's theory.

Of course, any chemical experiment can be criticised if impure substances are used. Lakatos assumes that we can test if a substance is chemically pure by applying complicated chemical purifying procedures to it. Let us with Lakatos assume the following "interpretative theory": If the chemical purifying processes *A* are applied to a gas, what remains will be pure chlorine. Lakatos thinks that by a refutation we do not have to question the "tested theory" (here Prout's theory), but can also question the interpretative theory. However, what Lakatos calls an interpretative theory is an auxiliary hypothesis used in the critical discussion of a theory. Lakatos thinks that Prout's theory was saved because a problematic auxiliary hypothesis assumed by Stas was questioned. But, as we have seen above, this way of understanding the resurrection of Prout's theory is historically mistaken: rather than an auxiliary hypothesis it was Prout's theory that was questioned and later modified.

How can the test statement that a substance is chemically pure be criticised, for example the test statement that a gas consists of chemically pure chlorine? The simplest way would be to show that the gas has some property which chemically pure chlorine does not have. (In this way Priestley and Lavoisier criticised statements that claimed that a gas was carbon dioxide or pure air. See chap. 6, B 2.3, above.) In a similar way Lakatos's imagined "interpretative theory" could be criticised by showing that a gas is not pure chlorine, in spite of the fact that the purifying processes *A* have been applied to it.

Test statements can be criticised by their deductive consequences, as Popper suggested. Can test statements be criticised indirectly, by their interpretative theories, as Lakatos suggests? Interpretative theories can be used as implicit auxiliary hypotheses when we discuss test statements. If we question an interpretative theory, this can have consequences for test statements. But interpretative theories are criticised with other test statements. As shown above, problematic test statements can also be criticised with other unproblematic test statements. Therefore, the "appeal procedure" pointed out by Popper, to criticise test statements by their deductive consequences, is the most basic one. If Lakatos's "appeal procedure" is intended to be the only way to criticise test statements, then this procedure is circular, since Lakatos uses test statements (empirical content) to judge interpretative theories (see chapter 4, C, above).

Lakatos has not given any satisfactory solution to the problem of how to criticise test statements. He does not seem to be quite satisfied with his ideas on test statements:

> The difficulties concerning the empirical basis which confronted 'naive' falsificationism cannot be avoided by 'sophisticated' falsificationism ei-

> ther. ... We cannot get rid of the problem of the 'empirical basis', if we want to learn from experience (Lakatos 1970, 131).

However, the problem is to show how we can learn from experience, how experience can be the 'impartial arbiter' of scientific controversy. Lakatos does not solve this problem.

Lakatos does not deal with the problem of test statements when he discusses his methodology of scientific research programmes. It is natural to regard interpretative theories as parts of research programmes. This leads to the problem of incommensurability, since you can argue that every research programme has its own interpretative theories. Lakatos does not deal with these problems in his discussions of the methodology of scientific research programmes. Therefore it is highly problematic for Lakatos how scientific research programmes can be judged and criticised empirically.

D. FEYERABEND ON TEST STATEMENTS

Like Kuhn, Feyerabend accepts the thesis of incommensurability and the gestalt psychological model of experience. When I discussed Kuhn's ideas on test statements, I criticised this model. I do not want to repeat this criticism, but will only deal with Feyerabend's most important examples of incommensurable theories here.

1. *Test Statements on the Tower Experiment*

Feyerabend maintains that the Ptolemaic and Copernican theories are dynamically incommensurable and that Galileo invented a new type of experience with metaphysical components (above, chap. 5, C 3.2). What new type of experience does Feyerabend think that Galileo invented?

Galileo and his Aristotelian adversaries agreed that a stone falls vertically from a tower. Thus they agreed on unproblematic test statements, and there was no incommensurability. However, Feyerabend maintains that Galileo introduced a new "natural interpretation" of the tower experiment, according to which all changes of motion presuppose a moving force (only relative motion is operative). Those adhering to the Ptolemaic theory, on the other hand, stuck to the Aristotelian "natural interpretation" of the tower experiment according to which all motion presupposes a moving force (all motion is operative).

These "natural interpretations" are auxiliary hypotheses used to derive prognoses about the tower experiment. With the Aristotelian auxiliary hypothesis you can derive that a stone falls obliquely from a tower if the earth rotates around its axis; with the Galilean auxiliary hypothesis you can derive that a stone falls vertically under the same conditions. In the next chapter I will discuss the methodological problems of such auxiliary hypotheses.

However, Feyerabend thinks that these auxiliary hypotheses were implicit and determined the interpretation of the experiment. Understood in this way, an adherent of the Ptolemaic theory immediately interprets the tower experiment as showing that the earth does not move around its axis. This would be the Aristotelian "natural interpretation" of the experiment. What would the Galilean "natural interpretation" of the experiment be? According to Galilean physics the tower experiment is compatible with the hypothesis of the (diurnal) motion of the earth: whether the earth rotates around its axis or not, the stone falls vertically from the tower. Therefore, the most natural Galilean interpretation of the tower experiment is that it leaves it open whether the earth rotates around its axis or not.

Feyerabend regards the "change of experience" in the following way: "An experience which partly *contradicts* the idea of the motion of the earth is turned into an experience that *confirms* it" (Feyerabend 1988, 73). He thinks that the situation is best described as a "*change of experience* that allows us to accommodate the Copernican doctrine" (Feyerabend 1988, 74) Can the tower experiment be interpreted as showing that the earth moves? The Galilean interpretation of the experiment is compatible with the idea of the motion of the earth as well as with the idea of a resting earth. "'Experience', i.e. the totality of all facts from all domains, cannot force us to carry out the change which Galileo wants us to introduce. The motive must come from a different source" (Feyerabend 1988, 73–74). Feyerabend thinks that this source is Galileo's wish to support the Copernican theory. In order to support this theory it is necessary to introduce the assertion that the earth moves, an assertion that is "absurd and counterinductive" from the Aristotelian point of view. Therefore Feyerabend thinks that

> arguments will not suffice—an interesting and highly important limitation of rationalism—and Galileo's utterances are indeed arguments in appearance only. For Galileo uses *propaganda*. He uses *psychological tricks* in addition to whatever intellectual reasons he has to offer (Feyerabend 1988, 67).

In order to achieve a real "change of experience," let us assume that the tower experiment can be interpreted as showing directly that the earth moves, at least if we use propaganda and psychological tricks. Then we would have two different natural interpretations of the tower experiment: an Aristotelian one saying that the earth is at rest, and a Galilean one saying that the earth moves. The Ptolemaic and Copernican theories are incommensurable, if the tower experiment can be interpreted in two fundamentally different ways.

Galileo and his opponents did not discuss the tower experiment in this way. For them the motion of the earth was not "observable," not even with natural interpretations, but a hypothesis to be tested. But in order to discuss the problem of incommensurability, I will assume that the tower experiment can be interpreted as showing that the earth moves and in this sense was "observable."

I assume that the Copernican and Ptolemaic theories were dynamically incommensurable in the sense that adherents of the two theories naturally interpreted the tower experiment in two different ways, with Galilean and with Aristotelian physics, and hence maintained the two different test statements: "At *k* the earth is at rest" and "At *k* the earth moves." Since the tower experiment can be "interpreted" and, perhaps, even "seen" in two different ways, an intersubjective agreement on unproblematic test statements is not possible. Earlier I argued that in such cases a criticist philosophy of science recommends that we should try to derive new and unproblematic test statements on which intersubjective agreement is possible. In this case such a derivation is easily done. If we introduce the Aristotelian theory of motion as an explicit auxiliary hypothesis, we can derive that a stone falls vertically from a tower if the earth moves:

> At *k* a stone falls freely from a tower. (Unproblematic initial condition)
> At *k* the earth is at rest. (Problematic initial condition)
> All motions presuppose a moving force. (Aristotelian auxiliary hypothesis)
> ___
> At *k* a stone falls obliquely down. (Unproblematic prognosis)

With the Galilean theory of motion another prognosis follows:

> At *k* a stone falls freely from a tower. (Unproblematic initial condition)
> At *k* the earth moves. (Problematic initial condition)
> All changes of motion presuppose a moving force. (Galilean auxiliary hypothesis)
> ___
> At *k* a stone falls vertically down. (Unproblematic prognosis)

These derivations show how the alleged incommensurability can be solved: from problematic test statements we can derive unproblematic ones with the help of auxiliary hypotheses. The new test statements are unproblematic and "commensurable." By observation it can easily be decided how a stone falls from a tower.

Galileo uses the tower experiment to test the hypothesis of the (diurnal) motion of the earth. From a historical point of view the diurnal motion of the earth was not an "observed effect," but a tested hypothesis. The tower experiment was problematic because the auxiliary hypotheses used to derive consequences from it were problematic, not because the test statements about the experiment were problematic. The test statement that a stone falls vertically down from a tower is unproblematic and intersubjective agreement can be reached about it. Problematic is which auxiliary hypotheses to use to derive consequences from the experiment. There was no intersubjective agreement on which auxiliary dynamic hypothesis to use, the Aristotelian or the Galilean. Later (chap. 7, E 2, below) I will discuss the tower experiment from this more interesting point of view.

2. *Observations with the Telescope and with the Naked Eye*

In the Ptolemaic and Copernican theories it is assumed that the distances between the earth and the planets vary. Hence both theories predict that the apparent size and brightness of the planets will vary when they are observed from the earth. With the naked eye such variations can be observed, but they did not agree with the predicted values. Therefore both theories could be regarded as falsified (above chap. 5, C 3.1).

When Galileo observed the variations in apparent size and brightness of the planets (especially of Venus and Mars) with the telescope he had constructed, he found that his observations agreed with the predictions of the Copernican theory and disagreed with those of the Ptolemaic theory. Thus observations of the brightness and size of the planets with the telescope gave other values than with the naked eye. Does this show that the Copernican and Ptolemaic theories were optically incommensurable, as Feyerabend argues?

It is necessary to separate two problems: (1) What can be seen with the telescope and with the naked eye? (2) Are observations with the telescope and with the naked eye reliable? The first problem is empirical and is related to test statements; the second problem is theoretical and is related to auxiliary hypotheses.

The test statements on the observations of the planets with the naked eye were unproblematic in the sense that intersubjective agreement could be reached about them. It is for example intersubjectively testable that Venus has almost constant size and brightness when observed with the naked eye. Different observers came to the same results. After a short period of initial problems, the telescopes were so good that intersubjective agreement was reached on the results of telescopic observations. Soon agreement was reached that the moons of Jupiter exist, that the apparent size of Venus varies widely and that Venus shows phases like the moon when observed with the telescope. What was problematic was not what can be seen with the telescope or with the naked eye, but which type of observation is reliable.

As long as the auxiliary hypotheses about the naked eye and about the telescope are implicit, the impression arises that the Ptolemaic and Copernican theories are incommensurable, that the Copernican astronomers accepted the observations with the telescope and the Ptolemaic astronomers accepted the observations with the naked eye. However, if the assumed auxiliary hypotheses are explicitly stated, they can be critically discussed and used to derive unproblematic test statements based on the observations with the telescope and with the naked eye.

Feyerabend's discussion of Galileo's use of the telescope does not show that the Ptolemaic and Copernican theories were optically incommensurable, as Feyerabend thinks, but that Galileo used other optical auxiliary hypotheses than the astronomers defending the Ptolemaic system. Feyerabend does not give any good reasons for believing that Galileo had to use propaganda, rhetoric, and psychological tricks in order to convince his opponents that the telescope is a reliable instrument. Instead we have a normal scientific discussion of different auxiliary hypotheses. I will deal with the methodological problems of such discussions in the next chapter.

Feyerabend's example is interesting mainly because it shows how important it is to make implicit auxiliary hypotheses explicit. Only then is it possible to discuss them critically and to test them empirically. What Irving Copi (1986, 511) says in another context is relevant for Feyerabend's discussion of Galileo's use of the telescope:

> Perhaps the most significant lesson to be learned from the preceding discussion is the importance to scientific progress of dragging "hidden assumptions" into the open. ... Because they are hidden, there is no chance to examine such assumptions critically and to decide intelligently whether they are true or false. Progress is often achieved by formulating explicitly an assumption which had previously been hidden and then scrutinizing and rejecting it.

The auxiliary hypothesis that the size and brightness of the planets can be correctly observed with the naked eye is an example of such a hidden assumption, which was not critically discussed before Galileo used the telescope and produced results different from the observations with the naked eye.

It is neither possible nor necessary to drag all hidden assumptions into the open, to make all implicit auxiliary hypotheses explicit. But it is important that it is always possible to make a problematic implicit hypothesis explicit and to discuss it critically. In actual research there are many implicit auxiliary hypotheses, as Duhem pointed out. When a scientist says that he measures an electrical resistance by observing the needle of an instrument, he implicitly assumes many auxiliary hypotheses. If a layman should ask how this is possible, most scientists would recommend a course in electricity, as Duhem says (1974, 144). Without doubt, this would make it easier for the laymen to understand the implicit auxiliary hypothesis. It is, however, important that the implicit auxiliary hypotheses can always be explicitly stated and critically discussed, especially when the auxiliary hypotheses are problematic, as the auxiliary hypotheses about the observations with the telescope and with the naked eye were when Galileo discussed them.

Today most astronomers would not explicitly state that the telescope is a reliable instrument, but would regard it as a part of the unproblematic background knowledge. However, if somebody should question the reliability of the telescope, it is possible to state the assumed auxiliary hypotheses explicitly. When Galileo first used the telescope for astronomical observations, the situation was different. It was not at all clear that the observations with the new instrument were reliable. In such a situation it is important to state the auxiliary hypotheses explicitly in order to be able to discuss them critically.

CHAPTER SEVEN

AUXILIARY HYPOTHESES AND THEORETICAL MODIFICATIONS

A. POPPER ON THEORETICAL MODIFICATIONS

Universal scientific hypotheses can be falsified by singular test statements (above chap. 2, C). However, this only shows that a falsificationist methodology is logically possible. Is it also factually possible? Formal logic alone cannot solve the methodological problems of science, such as: What test statements can be asserted in order to test scientific theories? What modifications of falsified theories are permitted? First when such questions are answered, can a falsificationist methodology be used in practice.

Conventionalist philosophers of science maintain that test statements are impregnated with theories (Duhem 1974, part II, chs. 4 and 6; Fleck 1981; Fleck 1983; Poincaré 1970, part III). For this reason they object to the positivist assumption that there is an infallible empirical basis. Popper also criticises the positivist assumption of an infallible empirical basis and thinks that "the philosophy of conventionalism deserves great credit for the way it has helped to clarify the relations between theory and experience" (Popper 1959, § 19, 80).

Kuhn, Lakatos, and Feyerabend regard the fallibility and theory-impregnation of test statements as an objection against falsificationism. In the previous chapter I showed how these methodological problems can be solved (cf. Andersson 1985).

Popper deals with some conventionalist objections to the idea that scientific theories are falsifiable. According to these objections any theory can be brought to agree to test statements in a number of ways: (1) by introduction of *ad hoc* hypotheses, (2) by change of basic definitions of the theory, (3) by doubting the reliability of the observer and especially the reliability of the test statements threatening the theory, (4) by doubting the acumen of the theoretician who is not able to explain the test statements with the available theory (Popper 1959, § 19, 81).

Popper thinks that these objections of an imaginary conventionalist are incontestable because it is impossible to decide by logical analysis alone if a theoretical system is falsifiable:

> Yet this only shows that my criterion of demarcation cannot be applied immediately to a *system of statements*... . The question whether a given *system* should as such be regarded as a conventionalist or an empirical one is therefore misconceived. *Only with reference to the methods applied* to a theoretical system is it at all possible to ask whether we are dealing with a conventionalist or an empirical theory. The only way to avoid conventionalism is by taking a *decision*: the decision not to apply its methods (Popper 1959, § 19, 82 (Popper's italics)).

Above (chap. 2, C) I showed that systems of statements are falsifiable just in the same sense as isolated statements are. From a logical point of view both theories and isolated hypotheses can be falsified. But of course a dogmatic scientist does not have to take criticism and falsificationism seriously, he can try to avoid refutations by immunising the theory against criticism, for example by using "conventionalist stratagems." This does not show that the theory is unfalsifiable, but that the scientist is dogmatic. Expressed in another way, you can deal with a scientific theory in an unscientific way.

Although Popper thinks that the conventionalist objections are incontestable, he formulates some methodological rules to prevent conventionalist stratagems. Against the first conventionalist objection, that the introduction of *ad hoc* hypotheses can always immunise theories against criticism, Popper suggests the following anti-conventionalist stratagem:

> As regards *auxiliary hypotheses* we propose to lay down the rule that only those are acceptable whose introduction does not diminish the degree of falsifiability or testability of the system in question, but, on the contrary, increases it. ... If the degree of falsifiability is increased, then introducing the hypothesis has actually strengthened the theory: the system now rules out more than it did previously: it prohibits more. We can also put it like this. The introduction of an auxiliary hypothesis should always be regarded as an attempt to construct a new system; and this new system should then always be judged on the issue of whether it would, if adopted, constitute a real advance in our knowledge of the world (Popper 1959, § 20, 82–83).

Here *ad hoc* hypotheses are regarded as special types of auxiliary hypotheses. In order to prevent immunisation of criticism with *ad hoc* hypotheses Popper rejects those auxiliary hypotheses that do not increase the degree of testability of the theoretical system. In this way Popper evaluates new auxiliary hypotheses. But he does not say that new theories should be evaluated in this way.

Popper further suggests that the introduction of a new auxiliary hypothesis should be regarded as an attempt to construct a new theory. According to Popper's (1979, 379) thesis against conventionalism, a theoretical system is closed in the sense that the introduction of new hypotheses into the theoretical system should be judged as an attempt to construct a new theory. The new theory must be evaluated anew and compared with the old one. A higher degree of testability is one form of possible progress, but not the only one. Another form of progress is to replace a falsified theory with a corroborated one.

Popper (1974) introduces another idea on *ad hoc* hypotheses in "Replies to my critics." There he regards those hypotheses as *ad hoc* that are introduced to explain a particular difficulty, but which are not independently testable. As an example Popper mentions Pauli's introduction of the hypothesis of the neutrino: it was introduced to explain a special effect (beta-decay), but could not be tested independently of it, that is, the only way to test the neutrino-hypothesis was to continue to observe beta-decay. On the other hand, in the beginning of the 19th century Leverrier and Adams introduced the hypothesis that an unknown planet exists in order to explain some irregularities in the movement of Uranus, the most distant planet known at that time. This hypothesis could be tested independently of the movement of Uranus when Leverrier and Adams calculated the path of the hypothetical planet. These predictions were tested, which led to the discovery of an unknown planet, later called Neptune (Popper 1974, 986).

In his later works Popper does not introduce any methodological rules against *ad hoc* hypotheses, but only says that in general our criterion of testability warns us against them (Popper 1974, 987).

The second conventionalist objection asserts that it is always possible to change basic definitions in order to evade falsificiations. Popper thinks that changes of definitions are permissible if useful, but that "they must be regarded as modification of the system, which thereafter has to be re-examined as if it were new" (Popper 1959, § 20, 83). When a falsification leads to changes of basic definitions in a theory, the theory is not kept unchanged in spite of the falsification, but changed due to the falsification, that is, an attempt has been made to learn from the falsification. (This anti-conventionalist rule is relevant also for a discussion of Stegmüller's non-statement view of scientific theories: changes of intended interpretations of a theory have a similar function as changes of basic definitions (cf. Stegmüller 1973, ch. 9, § 4)).

Against the third conventionalist objection, that test statements can always be doubted, Popper argues that intersubjectively testable and reproducible test statements cannot be doubted without further ado: "intersubjectively testable experiments are either to be accepted, or to be rejected in the light of counter-evidence" (Popper 1959, § 19, 84).

In *Die beiden Grundprobleme* (1979, 380–81) Popper deals with the conventionalist objections in another way. He calls hypotheses questioning (singular) initial conditions *ad hoc* hypotheses. In order to avoid theories becoming immunised against criticism by the introduction of ever new *ad hoc* hypotheses, Popper requires that initial conditions are only questioned in specific cases, not generally. (Otherwise a hypothesis like 'all swans are white' can always be saved by saying that all potential falsifiers are not swans.) If we require that test statements should be about reproducible and observable effects, as Popper does, this anti-conventionalist stratagem is not necessary.

In the methodological discussion it has remained unclear what sort of hypotheses are *ad hoc*. Therefore Popper's definition (1979, 380–81) is interesting. By doubting that initial conditions are satisfied, all falsifications of theories can be evaded. However, such doubts are interesting only if they are accompanied by new and reproducible counter-experiments.

The fourth conventionalist objection is unimportant. The mere belief that problems can be solved with a special theory if scientists are clever enough must not be taken seriously: the bare appeal to future explanations or discoveries can be disregarded (Popper 1959, § 20, 84). Of course scientists are allowed to believe that problems can be solved with a special theory. But such beliefs are always fallible, and there cannot be any guarantee of success.

B. Discussion of Popper's View

Popper argues against a conventionalist interpretation of scientific theories and against conventionalist stratagems. These arguments show that there are other ways of avoiding conventionalism than by merely taking a decision. Popper does not reject the conventionalist objections until after a critical discussion of them.

Are methodological decisions necessary in order to reject conventionalism? According to the second conventionalist objection against falsificationism it is always possible to avoid falsificiations of a theoretical system by changing basic definitions in the system. Popper argues that such changes are changes of the theoretical system. No methodological decision is necessary to understand this. The fact that the theoretical system is changed shows that the second conventionalist objection is false. If basic definitions of a theoretical system are changed, then the theoretical system is not immunised against criticism, but changed because of criticism.

According to the third conventionalist objection it is always possible to doubt test statements. However, if test statements describe inter-subjectively testable and reproducible effects, this objection can be rejected. Reproducible effects cannot always be doubted, but theoretical reasons or counter-experiments have to be given.

The thesis of incommensurability leads to a special case of the third conventionalist objection: Test statements are doubted because there are other theoretical interpretations of observations. In this case it is not enough to refer to inter-subjective testability or reproducibility. The reproducibility depends on the chosen theoretical interpretation of experience. Kuhn, for example, argues that a believer in the phlogiston theory always sees air without phlogiston, whereas a believer in the oxygen theory always sees oxygen. Above (chap. 6, A 2.3–4) I showed that in such cases of incommensurability it is possible to deduce new and unproblematic test statements and hence to compare the two theories. Therefore the third conventionalist objection is without force in this form also. Apparently incommensurable test statements can be made commensurable by deduction of new and unproblematic test statements.

The third conventionalist objection, that a sufficiently clever scientist can always solve problems with an available theory, can also be rejected with arguments. If the available theory is falsified and the test statements are not doubted, then the problem cannot be solved without changes of the available theory. Although it is possible that small changes are sufficient to solve the problem, the very fact that the theory is changed shows that the fourth conventionalist objection is false.

It is never possible to avoid a falsification merely by introducing new auxiliary (or *ad hoc*) hypotheses into a falsified theoretical system. If the theory T is falsified and modified by introducing the auxiliary hypotheses AH, then the theoretical system $(T \wedge AH)$ is also falsified (Pähler 1979, 199). In order to avoid the falsification, you have to change and modify the falsified theory. Therefore, no falsifications can be avoided by the introduction of new auxiliary hypotheses into the falsified theoretical system. In this case no methodological decisions are necessary to reject the first conventionalist objection.

This objection can be interpreted in another way: the auxiliary hypotheses are not introduced into the theoretical system, but used to doubt single falsifying test statements. This is a special case of the third conventionalist objection according to which test statements can always be doubted by the introduction of new auxiliary or *ad hoc* hypotheses. Lakatos raises this objection. The requirement that test statements should describe reproducible events can be used against it, as was indicated above.

All the conventionalist objections discussed by Popper can be criticised with logical and methodological arguments. It is possible to compare conventionalism and falsificationism critically and to reject the conventionalist objections against falsificationism.

Popper (1959, § 20, 82) thinks that the conventionalist objections are strong because his criterion of demarcation cannot be applied immediately to theoretical systems. But it is unimportant whether a theoretical system or an isolated hypothesis is tested empirically. If the theoretical system has empirical consequences, then it is falsifiable, and Popper's criterion of demarcation can be applied to it. Theoretical systems are just as falsifiable as isolated hypotheses. In both cases the falsifiability only depends on whether empirical consequences can be derived, that is whether there are any potential falsifiers. This can be decided by a logical analysis.

When a theoretical system is falsified, you do not know which parts of the system are false or how the falsified system should be modified. The falsification hits the whole system. Falsificationist methodology is no method of discovery and hence does not tell us how to modify falsified theories. You can only advance fallible conjectures on how to modify a falsified theory, and these conjectures have to be tested empirically. But this is also true for isolated hypotheses. If a hypothesis is falsified, it has to be changed, but you do not know how. Is a small modification of the falsified hypothesis sufficient or is a totally new hypothesis is necessary?

Popper (1959, § 20, 82–83) requires that the degree of falsifiability (testability) of a theoretical system must not be diminished by introduction of new auxiliary hypotheses into the system. This requirement is not necessary because it is not possible to diminish the degree of falsifiability of a theoretical system by introducing auxiliary hypotheses into it. If the auxiliary hypothesis AH is introduced into the theoretical system T, then the new system $T \wedge AH$ is at least as falsifiable as T (all potential falsifiers of T are also potential falsifiers of $T \wedge AH$). Thus the mere introduction of new auxiliary hypotheses cannot diminish the degree of falsifiability of a theoretical system or save a system from a falsification. No special anti-conventional stratagem is necessary in this case.

In the methodological discussion it has remained unclear what kinds of hypotheses are *ad hoc*. Hempel (1966, 30) remarks that with the benefit of hindsight "it seems easy to dismiss certain scientific suggestions of the past as *ad hoc* hypotheses, whereas it may be quite difficult to pass judgement on a hypothesis proposed in a contemporary context." According to Copi (1986, 509) it is agreed that *ad hoc* hypotheses should be rejected, but it is unclear what kinds of hypotheses are *ad hoc*. He discusses the following meanings of *ad hoc* hypotheses: (1) a hypothesis suggested only in order to explain a fact after the discovery of the fact, (2) a hypothesis that can only explain a single test statement or some single test statements, (3) a hypothesis that is a descriptive (or empirical) generalisation. But there is no reason to introduce special methodological rules against such hypotheses because (1′) all hypotheses are introduced in order to explain some special facts, (2′) no general hypothesis can explain only some single test statements. Hypotheses of the type "all ravens are black" are now and then called descriptive or empirical generalisations (although they are not found by generalisation). I do not think that there are any reasons to reject such low level hypotheses as *ad hoc*. Admittedly, a more general or theoretical hypothesis would be better. However, without such hypotheses we must be satisfied with a hypothesis that can explain a number of observed cases and that can be used to derive future predictions on ravens. Such "empirical generalisations" are falsifiable. From a falsificationist point of view there is no reason to reject them as *ad hoc*.

Popper (1979, 379–80) suggests that an *ad hoc* hypothesis is a hypothesis about some special initial condition. *Ad hoc* hypotheses in this sense are not introduced into the theoretical system, but used to eliminate special potential falsifiers. Logically, it is always possible to evade special falsificiations in this way. Such hypotheses are about special cases and are *ad hoc* in the etymological sense of the word. If such hypotheses are introduced *only* to evade falsificiations, they are to be rejected. Assume that the hypothesis 'all ravens are black' is falsified by the test statement 'at *k* there is a non-black raven'. Then it is always possible to evade the falsification by introducing *ad hoc* the hypothesis 'at *k* there is no raven'. If it is introduced only to save the tested hypothesis, it is of course inadmissible to proceed in this way. But if there are empirical (or other) reasons to think that there is no raven at *k*, then the hypothesis is interesting (even if *ad hoc*). Such criticism of the test statements is admissible and useful.

Popper (1974, 986) also defines an *ad hoc* hypothesis as a hypothesis that is introduced to explain some special types of facts, but that cannot be tested independently of these facts. As I argued above, all hypotheses are introduced to explain some special types of facts. It is an advantage if new hypotheses can be tested independently of the types of facts they were introduced to explain. However, hypotheses that are not testable in this way must not be rejected as inadmissible or *ad hoc*. Let us assume that Pauli introduced the hypothesis that there are neutrinos only in order to explain beta decay. The introduction of the neutrino hypothesis modified the theoretical system used to explain the mass-energy loss of a nucleus due to beta decay. The new theoretical system was falsifiable: it was not certain that future cases of beta decay would corroborate it. Since it was corroborated in such tests, I cannot understand why the neutrino hypothesis should be rejected as *ad hoc* or inadmissible. In a falsificationist methodology you can only require that falsified theoretical systems are changed, that theories are not immunised against criticism. The introduction of the neutrino hypothesis changed a theoretical system. The system remained testable, although it was not independently testable (in Popper's sense). In the context of discovery perhaps not everything, but much is allowed, even the introduction of hypotheses that are not independently testable.

Popper discusses three requirements of theoretical modifications: (1) they should not diminish the degree of falsifiability of the theoretical system, (2) they should be regarded as an attempt to construct a new theory, (3) they should be independently testable. In order to answer the conventionalist objections, it is sufficient to regard theoretical modifications as attempts to construct new theories. A high degree of falsifiability and independent testability are desirable, but not necessary. Competing theories are also desirable and should not be rejected rashly, for example neither because they have lower degrees of falsifiability nor because they are not independently testable. If a theory is corroborated, it should be regarded as a possible explanation, even when it is less falsifiable than a competing theory or not independently testable.

C. KUHN ON THEORETICAL MODIFICATIONS

Kuhn thinks that according to a falsificationist methodology falsified theories have to be totally eliminated and replaced by new ones, that is, that falsifications have to lead to scientific revolutions. As Kuhn correctly argues, in the history of science falsified theories are not always totally rejected, but often more or less extensively modified. Kuhn calls the research strategy consisting of modifications of an accepted theory normal science. He thinks that it shows that falsifications are not taken seriously and that hence falsificationist methodology is unrealistic (above, chap. 3, B).

Must a falsificationist methodology demand that falsified theories are totally rejected? Are modifications of falsified theories attempts to immunise them against criticism?

In most cases scientists do not test isolated hypotheses, but whole theoretical systems consisting of several hypotheses. Let us assume that we need n initial conditions and r general hypotheses in order to derive the prognosis P. As shown above (chap. 2, C 2, (15)), then the following falsification is valid:

(1) $C_1, \ldots, C_n, \neg P \vdash \neg(H_1 \wedge \ldots \wedge H_r)$

The whole theory is falsified, that is, not all hypotheses of the system can be true. The falsification does not tell us which parts of the system are false. According to De Morgan's law $\neg(H_1 \wedge \ldots \wedge H_r)$ is logically equivalent to $(\neg H_1 \vee \ldots \vee \neg H_r)$. Hence the falsification above is equivalent to:

(2) $C_1, \ldots, C_n, \neg P \vdash \neg H_1 \vee \ldots \vee \neg H_r$

A falsification only shows that at least one of the constituent hypotheses of a theoretical system is false, but it does not tell us which one. Therefore a falsification does not hit any specific part of a theoretical system. You do not know which part is false and has to be changed. However, it is not unclear whether a theory is falsified or not, only whether a specific part of it is falsified.

When a theoretical system is falsified, it has to be changed in order to be compatible with known and accepted test statements. Apart from agreement with accepted test statements, it is totally open how a falsified theory should be changed, if a small theoretical modification or a total rejection of the theory is necessary.

According to Duhem (1974, 216) a falsification only shows that a theory has to be changed, but it does not tell us how. The scientists have to use their intuition and acumen in order to find out how a falsified theory should be changed. Such changes can be made in many different ways:

> One [physicist] may be obliged to safeguard certain fundamental hypotheses ... The next physicist ... may decide to change some one of the essential assumptions supporting the entire system. The first physicist does not have the right to condemn in advance the boldness of the second one, nor does the latter have the right to treat the timidity of the first physicist as absurd. The methods they follow are justifiable only by experiment, and if they both succeed in satisfying the requirements of experiment each is logically permitted to declare himself content with the work that he has accomplished (Duhem 1974, 216–17).

When a theory is falsified, many different research strategies are possible. The scientists can try to solve the problem by small modifications or by total rejection of the falsified theory. Only empirical tests can decide which strategy is acceptable, not any methodological rules or decisions used *a priori.*

When a theoretical system is falsified, a criticist methodology only tells us that the system has to be changed, it does not tell us how. Often a falsified system can be modified in many different ways: some unimportant peripheral hypotheses or some important central hypotheses may be changed (cf. Kuhn 1957, 59–76 on the Ptolemaic theory). In a fallibilist methodology it is desirable that different research strategies are tried out, both conservative and revolutionary ones. From a fallibilist point of view the question of which strategy is the best one is totally open. (It is not a methodological, but an empirical question.)

Duhem (1974, 217) writes that logic alone cannot decide which research strategy to use. But his references to the "reasons which reason does not know," to "the mind of finesse," and to common sense do not solve the problem.

Kuhn tries to solve the same problem by reference to what scientists believe: if they believe in a paradigm, they use the cautious normal research strategy; if not, they use the daring revolutionary strategy (above chap. 3, B). Kuhn's attempt to solve the problem in a subjective and psychological way is unsatisfactory. Uncritical belief in theories is neither desirable nor necessary to explain "normal science."

When a theory is falsified, several research strategies are possible, daring "extra-ordinary" ones and cautious "normal" ones. Since we do not have any method of discovery, it is impossible to say *a priori* which strategy should be used: neither the intuition of the scientist, nor his belief, nor any methodological principles can solve this problem. The strategy actually chosen by the scientist is conjectural and fallible. Although the solution might be ingenious, although scientists might believe in it, this does not guarantee that the solution is acceptable. Only subsequent empirical testing can show this.

The research strategy of small theoretical modifications must not be dogmatic or accompanied by belief in a ruling theory (paradigm). Small theoretical modifications must not depend on blind belief in a generally accepted paradigm, but can be fallible conjectures. Scientific revolutions do not presuppose any crisis of belief, not even a falsification of the generally accepted theory. From a logical and methodological point of view, the "normal" and the "extra-ordinary" research strategies can always be used.

What does it mean that falsified theories should be rejected and replaced by new ones? In a scientific revolution it is obvious that a theory is rejected and replaced by another one. It is, for example, obvious that the geocentric theory of the planetary system and the phlogiston theory have been rejected and replaced by the heliocentric theory and the oxygen theory. When a theoretical system is modified, it is not so obvious that a version of the theory has been rejected and replaced by another one. But from a logical and methodological point of view this is the case. When a theory is slightly modified, then this modification represents a change in the theory and is an attempt to construct a new one. If an earlier version of a theory is falsified, then modifications of the theory show that the falsification is taken seriously, that attempts are made to learn from experience, and that the theory is not immunised from criticism.

For example, did Priestley try to immunise the phlogiston theory from criticism? He tried to solve problems by modifications of the phlogiston theory (Conant 1957, 90–104). These modifications show that he took empirical results seriously. A falsification of a special version of the phlogiston theory need not lead to the total rejection of the phlogiston theory as a research programme.

Even if the cautious or "normal" research strategy is often successful, it is not certain that problems can always be solved in this way (Grünbaum 1960, 75–78). If a central hypothesis of the theory is false, a bolder and more "extra-ordinary" strategy might be necessary.

In his criticism of falsificationism Kuhn presupposes that falsified theories must always be totally rejected. This presupposition is false. Falsificationism only requires that falsified theories are modified. Small modifications of theories do not show that falsificationism is unrealistic or that falsifications are not taken seriously, but that many research strategies are possible when a theory is falsified. In most cases we test whole theories and do not know which part of the theory is false when it is falsified. Duhem (1974, 187) saw clearly this systematic character of empirical tests:

> We have gone a long way from the conception of the experimental method arbitrarily held by persons unfamiliar with its actual functioning. People generally think that each one of the hypotheses employed in physics can be taken in isolation, [and] checked by experiment … . In reality, this is not the case.

D. LAKATOS ON THEORETICAL MODIFICATIONS

1. *Falsification and Elimination*

In his criticism of falsificationism Lakatos presupposes, like Kuhn, that falsified theories always have to be totally rejected. Lakatos (1970, 111) thinks that further work on a falsified theory is irrational. For what Lakatos calls the "savage falsificationist" falsifications are "methodologically conclusive." Thus, in this savage or "naive methodological" falsificationism every falsification must lead to a scientific revolution. Such a falsificationism would lead to permanent revolution in science.

Since there is no permanent revolution in science Lakatos thinks that falsificiations are unimportant and that the history of science refutes falsificationism (above chap. 4, B, last paragraph). This alleged refutation depends on the fact that in the development of science falsified theories are often modified and not totally eliminated. Therefore Lakatos rejects the "rule of elimination" that (logically) falsified theories should be eliminated and proposes a new rule according to which a theory or research programme should be eliminated if there is a better theory or research programme. Lakatos thinks that his new "rule of elimination" is rational and not merely an expression of subjective belief.

In the discussion of Kuhn's "normal science" (chap. 7, C above) I argued that the methodological consequence of a falsification is only that the falsified theory has to be modified, but that only subsequent empirical testing can show how extensive this modification has to be. From a logical point of view only a special version of the theory is falsified. Therefore it is wrong to think that falsified theories always have to be totally rejected (or eliminated), that it is irrational to continue to work on falsified theories or to modify them. Modifications of falsified theories are, even if small, changes of the theories and might be attempts to solve the problem posed by a falsification. If a theory is falsified, a small modification of the theory might very well be sufficient in order to get a corroborated theory. Since problems can be solved in this way, the research strategy consisting of small modifications of falsified theories is acceptable from a falsificationist and criticist point of view. Kuhn's and Lakatos's criticisms of falsificationism are logically and methodologically mistaken. The pragmatic or methodological meaning of a falsification is not that the falsified theory has to be totally rejected or eliminated. Such a "savage" or "naive" falsificationism is untenable for logical and historical reasons. The methodological consequence of a falsification is that the falsified theory has to be modified. *How* it has to be modified, to what degree, only empirical tests can show. There is no method of discovery, no heuristic and no methodology telling us how falsified theories are to be changed. We only have the method of conjectures and refutations, the hypothetico-deductive method of proposing fallible conjectures and testing them empirically. We have to learn from experience how falsified theories are to be changed.

In his discussion of "sophisticated methodological falsificationism" Lakatos (1970, 117) asks: "Why aim at falsification at any price? Why not rather impose certain standards on the theoretical adjustments by which one is allowed to save a theory?" Since Lakatos thinks that falsified theories have to be totally rejected, his question is understandable. Indeed, there are neither logical nor methodological reasons why falsified theories should be totally rejected, why scientists should stop working on them. In his sophisticated methodological falsificationism Lakatos tries to defend a research strategy consisting of small changes of falsified theories. However, since a correctly understood falsificationism allows this research strategy, Lakatos's proposals are superfluous.

In his methodology of scientific research programmes Lakatos tries to show that a research programme can progress in an ocean of anomalies (Lakatos 1970, 138), that anomalies (or falsifications) do not have to lead to the elimination of research programmes, but that the problems posed by anomalies can be solved (or assimilated) by development and slight modifications of the programme. This, however, is only surprising for a "naive" or "savage" falsificationist. If the logical aspects of falsifications are understood, nothing rules out that many problems can be solved just by small modifications of a theory or research programme. A falsificationist can only demand that falsifications are regarded as problems to be solved. But he has no universal method telling him how problems shall be solved, such as the unrealistic radical method of permanent scientific revolution.

Lakatos thinks that Popper recommends the total rejection of falsified theories. Is Lakatos's interpretation correct? According to Popper it is urgent to replace a falsified hypothesis by a better one (Popper 1959, § 22, 87, fn. 1). To replace a hypothesis need not mean to eliminate it totally. A falsified hypothesis can also be replaced by a slightly modified one. According to Popper (1959, § 82, 268) inter-subjectively testable falsificiations are in general regarded as final. But this only means that the falsification of a special version of the theory is generally regarded as final, not that every possible version of the theory is falsified, that the theory cannot be modified into a better one.

Popper (1965, 312, fn. 1) thinks that "the dogmatic attitude of sticking to a theory as long as possible is of considerable significance. Without it we could never find out what is in a theory—we should give the theory up before we had a real opportunity of finding out its strength." Lakatos finds Popper's approval of the dogmatic attitude strange and interprets it as a sign of an anomaly in Popper's methodology (Lakatos 1970, 177, fn. 3). Lakatos thinks that his new rules of elimination of theories and of research programmes can get rid of this anomaly.

To stick to a theory as long as possible can mean that falsifying test statements are questioned or that a falsified theory is slightly modified. However, such a defence of a theory need not be dogmatic, but can be a fallible conjecture on how to solve problems. Small theoretical modifications or criticism of test statements need not be expressions of dogmatism. Sticking to one research strategy under all circumstances, for example sticking to the "normal" and cautious strategy of defending accepted theories, is dogmatic. But it is just as dogmatic always to reject falsified theories totally, that is, to stick to an "extra-ordinary" and bold strategy of research under all circumstances.

Since there is no method of discovery, we have to use different research strategies in order to solve scientific problems. From a criticist and fallibilist point of view it is desirable that many different theories and strategies are tried. This pluralism of theories and of research strategies is a consequence of fallibilism and has a critical function. A criticist and fallibilist methodology should not reject cautious and unspectacular research strategies as dogmatic or uncritical.

2. *Normal Science and Heuristics*

Lakatos advances his methodology of scientific research programmes in order to be able to deal with the "remarkable continuity" of science. When the cautious research strategy of small theoretical modifications is used, the development of science is more or less continuous because central hypotheses of the theory are not questioned.

When the cautious strategy is used, falsificiations do not lead to totally new theories. This strategy plays an important role in the history of science: scientists seldom suggest totally new theories in order to solve problems, but take already available theories and solutions as their points of departure.

According to Duhem (1974, 200–21) hypotheses are not the product of sudden creation, but the result of a progressive evolution. No scientific theory has been created out of whole cloth, but is the result of a "series of retouchings which from almost formless first sketches have gradually led the system to more finished states."

The continuous development of science has led many philosophers of science to think that it is possible to find a method of scientific discovery. Norwood Russell Hanson, for example, criticises the hypothetico-deductive view of science for not telling us how new hypotheses are discovered. He does not find it sufficient to say the new hypotheses are discovered by creative imagination, since the proposal of a new hypothesis very often is a "reasonable affair" (Hanson 1958, 71).

Kuhn thinks that science develops continuously because scientists believe in paradigms, which govern the growth of science and have a heuristic function. Lakatos's scientific research programmes have similar functions.

Kuhn's and Lakatos's explanations of the continuous development of science are unsatisfactory. Kuhn thinks that scientists do not question the paradigm because they uncritically believe in it; Lakatos assumes that scientists decide to make central parts of a research programme unfalsifiable. Duhem, on the other hand, assumes neither uncritical belief nor methodological decisions, but treats the cautious and the bold research strategies as equal. From a fallibilist point of view, Duhem's pluralist conception is the only satisfactory one. In order to understand the continuous development of science neither uncritical belief nor methodological decisions are necessary. In order to follow a research programme, you do not have to believe in it uncritically or to make parts of it unfalsifiable, but you can regard it as consisting of fallible suggestions and hypotheses. When a theory is falsified, you do not have to reject it totally, but you can try to modify it by changing different parts of it, including peripheral and unimportant parts. In this way you can test the strength of a research programme or theory. In order to understand this research strategy, no kind of dogmatism has to be assumed, only ordinary fallibilism.

If a scientist adopts a research programme or a heuristic idea, this does not mean that he has a method of discovery. In order to work out a research programme or a heuristic idea or in order to modify a theory, the scientist has to use his creative imagination. More creative imagination is necessary in order to find new theories than in order to modify them. But the less spectacular task of modifying and improving available theories also presupposes creative imagination. In all areas great creativity and totally new ideas are rare: "Men with a presentiment of new truths are rare in all the sciences; most men develop and follow the ideas of a few others" (Bernard 1957, 34). Also the "normal" scientist, who follows a research programme or paradigm, has an important task in science and needs creative imagination.

3. *Falsifiability of Scientific Theories*

Lakatos maintains that the most admired scientific theories, like Newton's theory of gravitation, are unfalsifiable (Lakatos 1970, 100). First Lakatos thinks that scientific theories are tested together with *ceteris paribus* clauses. If the theory is falsified, we do not have to reject it, but can reject a *ceteris paribus* clause. Later Lakatos no longer thinks that *ceteris paribus* clauses are necessary to test a theory, but that whole theoretical systems are tested. If such a system is falsified, the problem arises: which part of the system is false, and which initial conditions, hypotheses, or auxiliary hypotheses should be changed? In his sophisticated methodological falsificationism Lakatos thinks that any part of a falsified system can be changed. He even suggests that any research strategy might be successful, if pursued with sufficient energy and creativity. Thus parts of a theoretical system can be immunised against criticism if other parts are modified with sufficient energy and creativity. In this way Lakatos expresses a conventionalist objection against falsificationism (as the "Quine-Duhem thesis"). In its weak interpretation this objection says that parts of a theoretical system cannot be falsified, in its strong interpretation that there is no rational "selection rule" among alternative theoretical systems (Lakatos 1970, 184–54).

Thus Lakatos discusses two objections against falsificationism: (1) the most admired scientific theories are unfalsifiable, (2) constituent hypotheses of a theoretical system are not falsifiable. He elucidates these objections by discussing Popper's example of a scientific explanation, the breaking of a thread, and by discussing Newton's theory of gravitation.

3.1. *The Breaking Thread*

Popper explained the breaking of a thread in the following way:

> Whenever a thread is loaded with a weight exceeding that which characterizes the tensile strength of the thread, then it will break. (Hypothesis_1)
> The weight characteristic for this thread is 1 *lb*. (Hypothesis_2)
> An iron weight of 2 *lbs* was put on the thread located in the space-time region *k*. (Initial condition)
> ---
> The thread located in the space-time region *k* breaks. (Prognosis)

Lakatos thinks that if there is a falsification, we may restore consistency by changing any part of the falsifying argument: the negation of the prognosis, the initial condition, or any of the hypotheses. Thus we may doubt that the thread did not break (negation of the prognosis), that an iron weight of 2 *lbs* was put on the thread (first part of the initial condition), or that an ordinary thread was used in the experiment (second part of the initial condition). We may also change any of the hypotheses and maintain that rather than the mass, the pulling force of the weight is relevant (modification of the first hypothesis), or we may maintain the tensile strength of the thread was not 1 *lb*, but more (modification of the second hypothesis) (Lakatos 1970, 185).

First I will discuss the possible modifications of the test statements (initial conditions and prognoses). Lakatos repeats the conventionalist objection that falsifying test statements can always be doubted. This objection is not true for test statements describing inter-subjectively testable and reproducible effects. If somebody should doubt that the thread was loaded with 2 *lbs* or that the thread broke, then he could examine the thread and the weight used in the experiment once again. If it should be impossible to test the experimental set-up in this way, he could repeat the falsifying experiment, perhaps with special control of disturbing factors. Lakatos argues that the scales might go wrong, that the observing assistant might be unreliable, and that a magnet might be hidden in the ceiling of the laboratory. If the experiment is reproduced after such doubts, it is a good idea to use other scales and assistants or to control them better, and to check if there are any disturbing magnetic forces in the laboratory. If closer examination shows that the scales really were wrong, the assistant unreliable, or that there were disturbing magnetic forces, then the criticism of the test statements was justified, otherwise not. In this way the test statements can be tested. Experience rules out many possible changes of test statements. Arbitrary modifications of test statements about reproducible experiments and inter-subjectively testable effects are not possible.

Lakatos suggests two theoretical modifications. The first of them is a slight modification: not the mass of the weight, but its pulling force is relevant. This modification is the reason why Lakatos introduces a second initial condition: "there was a hidden magnet (or hitherto unknown force) located in the laboratory ceiling" (Lakatos 1970, 185). This more or less occult force, magnetic or unknown, in the laboratory ceiling could be used to immunise criticism. In this case the corresponding *ad hoc* hypotheses are not dangerous, since the force pulling the thread can be tested independently, for example with a spring balance attached to the thread. There are also independent tests of magnetic forces. It is of course more difficult to test unknown forces, whether in the laboratory ceiling or in other places. If the postulated force is totally unknown and cannot be reproduced in future experiments, then we have a typical occult effect that we do not have to take seriously (Popper 1959, § 8).

Lakatos also suggests that the tensile strength of the thread depends on whether it is moist and modifies the second hypothesis to 'The tensile strength of the actual thread, since it got moist, was 2 *lbs*' (Lakatos 1970, 185). However, whether the tensile strength of threads really depends on their moisture can easily be tested in experiments with moist threads. It is not at all sufficient to propose modifications of falsified theories inspired by creative imagination; these modifications must also be corroborated. This is a sufficient restrictive "selection rule" of theories.

A closer examination of Lakatos's first example thus shows that it is not at all as easy to immunise hypotheses against criticism as Lakatos thinks. In science the creative imagination is limited by empirical tests. Therefore you must not modify test statements or hypotheses arbitrarily.

3.2. *Newton's Theory of Gravitation*

Lakatos thinks that such a much admired theory as Newton's theory of gravitation is unfalsifiable. If the theory is falsified, it is always possible to introduce auxiliary hypotheses in order to immunise it from criticism.

Newton's theory of gravitation can be tested by observing whether the planets move according to the predictions of the theory. For such tests we need hypotheses about the structure of the solar system. Without such hypotheses about the relevant gravitational forces, it is impossible to derive any predictions about the movements of the planets. When we test Newton's theory of gravitation, we test it together with an auxiliary hypothesis on the structure of the solar system. If a planet does not move according to our predictions, the whole theoretical system is falsified. Then we only know that Newton's theory or the auxiliary hypothesis about the solar system is falsified. The alleged unfalsifiability of Newton's theory depends on the fact that in this special experiment Newton's theory is tested together with an auxiliary hypothesis.

Lakatos's example only shows that you cannot test the isolated Newtonian theory by observing the movements of the planets. If there is a falsification, it is possible that the auxiliary hypothesis on the structure of the solar system is false. When confronted with the anomalous movements of Uranus, Leverrier (and Adams) assumed that the auxiliary hypothesis is false, that there must be an unknown planet in the solar system disturbing the movement of Uranus. Their calculations of the path of this planet made their hypothesis independently testable and led to Galle's discovery of Neptune in 1846 (Grosser 1979, ch. 7; Wolf 1890–92, vol. 2, § 559). Leverrier's and Adams's hypothesis was testable independently of the movement of Uranus, since it allowed predictions of the positions of the unknown planet. Galle and other astronomers tested these predictions and found an unknown planet in the predicted positions. Leverrier and Adams succeeded in defending Newton's theory because their auxiliary hypothesis on the solar system was corroborated.

Lakatos, however, thinks that their defence would also have been successful without corroboration of the auxiliary hypothesis. He thinks that it is always possible to introduce additional auxiliary hypotheses on the brightness of the hypothetical planet or of other disturbing factors, if the original auxiliary hypothesis should not be corroborated.

Lakatos's opinion is problematic from a historical point of view. Without doubt, Leverrier was a scientist with much creative imagination. He tried to use it in order to explain small, but inter-subjectively testable anomalies in the movement of the planet Mercury. First he used an auxiliary hypothesis similar to the one above and assumed that there was an unknown planet between Mercury and the sun. As in the earlier case, he calculated the path of the unknown planet and even named it 'Vulcan'. Although the anomalous movement of Mercury could be explained in this way, Leverrier's second hypothesis was not accepted because there were no reproducible observations of the hypothetical Vulcan (Hanson 1962, 375). The additional auxiliary hypotheses suggested by Lakatos, low brightness of Vulcan or disturbing "cosmic clouds," were not suggested by Leverrier (Hanson 1962, 365–77). Are there any methodological reasons for this?

Lakatos's additional auxiliary hypotheses are independently testable and not very plausible. The anomalous movements of Uranus and Mercury were reproducible effects. Leverrier first had to investigate whether these anomalies could be explained by assuming that there is an unknown disturbing planet. Many possible anomalies in the movements of planets cannot be explained in this way. When the path of the hypothetical planet has been calculated, arbitrary additional auxiliary hypotheses are no longer possible because the size and the brightness of the assumed planet can also be calculated. Therefore, scientists cannot assume arbitrarily, as Lakatos suggests, that the hypothetical planet is so small that it cannot be observed with the best available telescopes.

Lakatos's other additional auxiliary hypothesis, that the hypothetical planet is hidden by an unknown "cosmic cloud," is also independently testable. This hypothesis must explain why the hypothetical planet with calculated path and mass is never observed. It must be assumed that the cosmic cloud moves in such a way that it always hides the unknown planet: a not very plausible assumption. Furthermore, the assumed cosmic cloud must not only hide the hypothetical planet, but also stars in its neighbourhood, which can be tested independently.

The independent testability of Lakatos's suggested additional auxiliary hypotheses makes their chances of success small. In the discussion of the planet Vulcan no additional auxiliary hypotheses on cosmic clouds in the solar system were suggested, probably because such hypotheses are not easy to reconcile with the astronomical background knowledge. Furthermore, a cosmic cloud in the solar system that is so dense that it absorbs the light of a planet would already have been known to us.

Lakatos does not succeed in showing that the Newtonian theory is unfalsifiable. The discussion of the anomalous movements of Uranus and Mercury only shows that the Newtonian system was tested together with auxiliary hypotheses about the structure of the solar system. These anomalies falsify a theoretical system consisting of the Newtonian theory and a hypothesis about the solar system. However, the discussion does not show that the Newtonian theory was unfalsifiable or that the introduction of new auxiliary hypotheses could always save the theory from refutations.

If the testing of theoretical systems is properly understood, no criticism of falsificationism can be based on Lakatos's discussion of the Newtonian theory. Leverrier's strategy to change the auxiliary hypothesis about the solar system agrees with a falsificationist methodology. It did not involve immunisation against criticism, but in changing a part of the falsified theoretical system. There is no guarantee that such a strategy will always succeed. Empirical testing showed that Leverrier's strategy was successful the first time he used it, but not the second time: Neptune was discovered, but not Vulcan.

Since Vulcan was not discovered, Leverrier suggested a number of other changes of the hypothesis of the solar system, for example that there were unknown asteroids between the sun and Mercury or that the mass of the sun was not uniformly distributed. These possible modifications were also independently testable and rejected (Hanson 1962, 375). Changes of the Newtonian theory were also suggested: "The most arresting hypothesis of the late nineteenth century was that of Asaph Hall, who returned to the conjecture that solar gravitation may not function exactly as the inverse square of the distance" (Hanson 1962, 376).

The discussion about the anomalous movement of Mercury shows that in this case small deviations were also taken seriously, that Lakatos is not right when he thinks that small deviations that cannot be explained by introduction of new auxiliary hypotheses are buried in the dusty volumes of periodicals and never mentioned again (Lakatos 1970, 101). On the contrary, Leverrier maintains that "no planet has exacted more pain and trouble of astronomers than Mercury, and has awarded them with so much anxiety and so many obstacles" (Hanson 1962, 365).

3.3. *The Duhem-Quine Thesis*

Lakatos thinks that it is always possible to defend a part of a theoretical system if another part is modified with sufficient creative imagination (Lakatos 1970, 184). He refers to Quine in order to support this view: "Any statement can be held true come what may, if we make drastic enough adjustments elsewhere in the system" (Quine 1961, 43). Lakatos also refers to Duhem. But Duhem only maintains that many hypotheses cannot be tested in isolation. He did not defend Lakatos's and Quine's view that any statement can be defended or "held true" with sufficient imagination. Duhem (1974, 215) asks what follows from the fact that many hypotheses cannot be tested alone:

> Does it follow that these hypotheses placed beyond the reach of direct experimental refutation have nothing more to fear from experiment? That they are guaranteed to remain immutable no matter what discoveries observation has in store for us? To pretend so would be a serious error.

Perhaps most isolated observations or test statements can be explained by a given theory if we are allowed to introduce arbitrary auxiliary hypotheses. However, if we require that reproducible effects are explained, as for example the anomalies in the movement of Mercury, it is not so easy to find an explanation with a given theory, not even if we are allowed to introduce auxiliary hypotheses arbitrarily. If we also require that the modified theoretical system shall be severely tested and corroborated, it is still more difficult to find a satisfactory explanation. With the Vulcan-hypothesis Leverrier could explain the movement of Mercury, but the hypothesis was not corroborated and hence not accepted. Thus empirical tests rule out many arbitrary modifications of theoretical systems.

According to Duhem (1974, part 2, ch. 6) most hypotheses are tested as parts of a theoretical system. Quine (1961, 41) goes further and maintains that "our statements about the external world face the tribunal of experience not individually but only as a corporate body." Lakatos shares this holistic view and does not see any reason why we should not say that each test is a challenge to the whole of our knowledge (Lakatos 1970, 186).

Hypotheses are not always tested as parts of theoretical systems. If an isolated hypothesis has empirical consequences, then the hypothesis can be tested in isolation and without the help of further hypotheses. In a test those hypotheses that are used as premises to derive the test statement are tested. These premises are different in different tests, as Quine later (1991) realised. Sometimes only an isolated hypothesis is tested, sometimes a system consisting of many hypotheses. It is not true that in all experiments only whole theoretical systems are tested, and it is a vast exaggeration to maintain, as Quine does, that in all experiments the whole of our knowledge is tested. This view is a holistic dogma (cf. Popper 1965, 239).

If hypotheses can be tested in isolation, then they should be tested in this way because such tests are severe tests. Otherwise hypotheses must be tested together with other hypotheses. There are no guarantees that hypotheses or theoretical systems are corroborated in tests or that logically possible modifications of hypotheses or theoretical systems survive empirical tests. Especially, it is not certain that a part of a theoretical system can be defended when other parts of the system are changed. Hence empirical tests are much more discriminating than Quine and Lakatos realise. When a theoretical system is refuted by a reproducible effect, it is not at all certain that some slight modifications of the system are sufficient to solve the problem. The history of science shows that sooner or later most theories are abandoned. If they could always be defended with sufficient imagination, this would be difficult to understand. The study of the history of science shows "that it would be very imprudent for us to say concerning a hypothesis commonly accepted today: 'We are certain that we shall never be led to abandon it because of a new experiment, no matter how precise it is'" (Duhem 1974, 212).

E. Feyerabend on Theoretical Modifications

In *Against Method* Feyerabend maintains that Galileo immunised the Copernican theory against criticism with *ad hoc* hypotheses. Observations with the naked eye of the variations in size of the planets did not agree with the Copernican theory (above chap. 5, C 3.1). The tower experiment refuted the hypothesis of the diurnal motion of the earth and thus the Copernican theory (above chap. 5, C 3.2). Galileo's observations of the variations in size of the planets with the telescope agreed with the Copernican theory. And with Galileo's new theory of motion the tower experiment no longer refuted the Copernican theory. But Feyerabend thinks that Galileo's new auxiliary hypotheses, the hypotheses that the telescope is a reliable instrument and his new theory of motion, were advanced *ad hoc* only in order to save the Copernican theory.

1. *Galileo's Observations with the Telescope*

Especially problematic for the Copernican (and Ptolemaic) theory were the small variations in size and brightness of the planet Venus. According to both theories the distance to Venus should vary considerably, but in spite of this the brightness and size of Venus hardly varied when observed with the naked eye. Does this show that the Copernican theory was refuted from its conception?

With the naked eye no reliable observations of the size of the planets are possible. Giordano Bruno discussed the problem and came to the conclusion that no observations with the naked eye of a radiant object allowed any inferences to its actual brightness or distance (Feyerabend 1975, 111). It was also unknown whether the planets radiate reflected light or light of their own. Without such auxiliary hypotheses the variations in brightness of the planets could not be used to test the astronomical theories. Since both test statements and auxiliary hypotheses were problematic, it is not so surprising that neither Ptolemy nor Copernicus tried to explain the variations in size and brightness of the planets exactly, but regarded these phenomena as open problems. At least it is clear that there were no unproblematic refutations of the theories.

Galileo's observations of the planets with the telescope showed how the open problem could be solved. In the telescope the planets appeared to be disklike. For the first time distinct observations of the sizes of the planets were possible. These observations agreed with the predictions derived from the Copernican theory. With the telescope Galileo discovered that the planets show phases like the moon. Especially varied were the phases of the inner planets Venus and Mercury. This allowed Galileo to explain the almost constant brightness of Venus when observed with the naked eye. When Venus is close to Earth, it is "new": only a small part of the illuminated surface can be seen. Therefore Venus is not as bright as could be expected when Venus is close to Earth. When Venus is far away from Earth, it is almost "full": almost the whole illuminated surface can be seen. The phases of Venus compensate for the variations in size and distance. This explains the almost constant brightness of Venus when observed with the naked eye (Price 1959, 197–218, especially 213). Galileo's observations with the telescope thus solved the problem of explaining the variations in size and brightness of the planets and the problem of whether the planets radiate reflected light or light of their own. Since the planets show phases, they must radiate reflected light.

Still more important was that the observations of the phases of Venus allowed a crucial experiment between the Copernican and Ptolemaic theory. The positions of the planets could be explained with both theories. Both theories predicted that the planets would show phases if they radiated reflected light. But they predict different phases. These differences are especially great for Venus. According to the Copernican theory Venus should appear almost "full" when far away from Earth; according to the Ptolemaic theory Venus should at that time be almost "new," that is, only a small part of the illuminated surface of Venus should be seen from Earth. With the telescope Galileo observed that Venus is almost "full" when far away from Earth. These observations agree with the Copernican theory and refute the Ptolemaic theory.

To this very important crucial experiment Feyerabend remarks that the phases of Venus could also be explained with the geocentric theory of the planetary system suggested by Tycho Brahe (Feyerabend 1988, 170, 254). Kuhn (1957, 224) argues in a similar way:

> The Tychonic system ... provides as good an explanation as the Copernican for the observed phases of and distance to Venus. Therefore, the telescope did not prove the validity of Copernicus' conceptual scheme. But it did provide an immensely effective weapon for the battle. It was not proof, but it was propaganda.

Like Duhem earlier, Feyerabend and Kuhn argue that a crucial experiment cannot prove that a hypothesis is true (Duhem 1974, part 2, ch. 6, § 3). But in a falsificationist methodology the function of a crucial experiment is not to prove, but to falsify. Of course the observations of the phases of Venus cannot prove that the Copernican theory is true. (No observations can do so.) But they falsified the Ptolemaic theory. This is important and shows that the observations of the phases of Venus were not only propaganda.

The observations with the telescope contributed to the solution of further problems. Copernicus assumed that the moon moved around the earth, but that the planets moved around the sun. Many astronomers thought that the hypothesis that only the moon moves around the earth was *ad hoc*. Galileo's discovery of the moons of Jupiter made the geocentric movement of the moon around the earth less strange: not only the earth, but also Jupiter had satellites. After the discovery of the moons of Jupiter it no longer seemed *ad hoc* to assume that the moon moves around the earth, while the planets move around the sun.

Tycho Brahe thought that the Copernican theory was falsified. His most important argument was the following. According to the Copernican theory the earth moves around the sun. If there is such a movement, the stars should be seen in different positions every six months. However, such a parallax cannot be observed. Copernicus discussed this problem, but said that it only shows that the stars are very far away from the earth (Berry 1961, 123–24). Copernicus thus regarded an auxiliary hypothesis about the distance to the stars as falsified. Tycho Brahe, however, regarded this hypothesis as independently testable. Observed with the naked eye the brighter stars are not punctiform, but appear disklike. According to Tycho Brahe this shows that they cannot be very far away (Dreyer 1953, 360–61). Therefore Galileo's observation that the fixed stars appear as mere luminous points in the telescope was very important. "This swept away the very serious objection raised by Tycho that a star having no annual parallax and yet showing a considerable apparent diameter must be incredibly large" (Dreyer 1953, 414).

Thus Galileo's observations with the telescope solved a number of open problems and agreed with the Copernican theory, and refuted the Ptolemaic theory. However, this presupposes that his observations were reliable. Feyerabend argues that in the beginning they were very problematic. He thinks that Galileo assumed that the telescope was reliable in spite of many problems mainly because it supported the Copernican theory. Those supporting the Ptolemaic theory assumed, with Aristotle, that only observations with the naked eye were reliable, and hence that all observations with instruments were unreliable. This is a discussion of different auxiliary hypotheses, the Galilean hypothesis that the telescope is reliable, and the Aristotelian hypothesis that only observations with the naked eye are reliable.

Galileo's auxiliary hypothesis on the reliability of the telescope was certainly fallible. Is Feyerabend right when he says that a rational discussion of this hypothesis or other auxiliary hypotheses is impossible?

The hypothesis that the telescope is reliable is independently testable: observations with the naked eye or with the telescope can be tested without any assumptions on the truth of the Copernican or Ptolemaic theories. And they were tested in this way by Galileo. From a falsificationist point of view, Galileo's criticism of the competing auxiliary hypothesis that observations with the naked eye are reliable is especially interesting. Galileo argues that a bright planet, for example Jupiter, appears to have an apparent diameter when observed in the middle of the night with the naked eye, but that the same planet appears to be much smaller when observed at dawn. The same is also true for the brighter fixed stars. Galileo explains these differences by assuming that small radiant objects are seen too large against a dark background. He points out how this optical illusion (irradiation) can be tested: if there is no dark background, there is no irradiation. For example, there is no irradiation if a bright star is observed at dawn (Galilei 1957, 45; Galilei 1967, 103; Koyré 1957, 91).

Thus Galileo not only criticises the hypothesis that observations with the naked eye are reliable, but he also suggests a new auxiliary hypothesis explaining why such observations are sometimes illusory. Galileo's new hypothesis (of irradiation) explains why the observation of the sizes of the planets with the naked eye agreed neither with the Ptolemaic nor with the Copernican theory. His new hypothesis also tells us when observations with the naked eye are unreliable.

The other auxiliary hypothesis, that observations with the telescope are reliable, was also independently testable. Distant objects on the earth could be observed with the telescope. Such observations showed that the telescope was reliable on the earth. Feyerabend objects that this does not show that the telescope was reliable when used in astronomy (Feyerabend 1988, 88). In order to show that the astronomical use of the telescope was problematic Feyerabend gives the following reasons. In a drawing of the moon in *Sidereus Nuncius* a crater is much too big (Feyerabend 1988, 97). In the beginning not all observers could see the moons of Jupiter in the telescope. It was problematic whether these observations were reproducible (Feyerabend 1988, 90–92). Galileo could not explain how his telescope functioned. And Kepler's explanation (in 1611) of the telescope could easily be falsified (above chap. 5, C 3.1). Do these problems show that Galileo's hypothesis that the telescope is reliable was as problematic as the hypothesis that observations with the naked eye are reliable?

Galileo's drawings of the moon in *Sidereus Nuncius* do not show that the telescope is unreliable. In this first publication of the observations with the telescope Galileo did not try to draw an exact map of the moon, but was only concerned with showing that there are mountains on the moon. This was an important discovery because Aristotle and his followers thought that the moon was perfectly spherical. Galileo writes that it was repeated observations that first led him to the view that there are mountains on the moon (Galilei 1957, 31). In order to refute the hypothesis of the perfectly spherical form of the moon, no exact maps of the moon were necessary. Therefore Galileo only draws one of the craters of the moon (too large) in order to show that the moon is not spherical. When Galileo's problem situation is understood, Feyerabend's somewhat pedantic objection, that Galileo's drawings of the moon were no exact maps, loses its force.

Galileo's first observations of the moons of Jupiter were not so easy to repeat. The first telescopes were primitive and the first observers untrained in using the new instrument (Dijksterhuis 1961, 382). In the beginning it was problematic whether Galileo's observations were reproducible. But with better instruments and trained observers agreement was soon reached that the moons of Jupiter really could be seen in the telescope. Already in 1610 Father Clavius, astronomer of the Jesuit Collegium Romanum, admitted that Jupiter had moons. Other astronomers soon agreed. Feyerabend explains this by saying that they were lax in their investigations (Feyerabend 1988, 94, fn. 22). Simpler than this *ad hoc* hypothesis is to assume that the moons of Jupiter were accepted so soon because they could be seen in the telescope. Galileo's hypothesis that Jupiter is surrounded by moons was intersubjectively testable.

Immediately after the publication of *Sidereus Nuncius*, where Galileo reports his observations with the telescope for the first time, Kepler wrote that it is necessary to explain how the telescope functions. Already in 1611 Kepler succeeded in giving such an explanation of the Galilean telescope and even in predicting how a better telescope can be built. This telescope was later built and generally used in astronomy.

Feyerabend argues that Kepler's explanation was unsatisfactory because Kepler's theory could easily be falsified: if an object is observed near the focus of a lens, then the object is not seen very far away, as could be expected with Kepler's theory. This shows that Kepler could not explain how objects are seen through lenses. But it does not falsify Kepler's theory as a theory on how light is refracted in lenses. As a theory on the refraction of light Kepler's theory could explain the telescope well. That Kepler's predictions on the refraction of light in lenses are correct can easily be tested independently. For example, according to Kepler's theory light rays from the focus of a lens are parallel after refraction (cf. Feyerabend 1988, 104).

Feyerabend argues that even if Kepler could explain the telescope optically, this does not prove that observations with the telescope are reliable. According to Feyerabend a new theory of vision was necessary (Feyerabend 1981b, 239). But such a theory would not have proved the reliability of the telescope either. The auxiliary hypothesis that the telescope is reliable was also inter-subjectively testable without a new theory of vision. This is enough for a rational discussion of the reliability of the telescope.

If we assume that the telescope is reliable, Galileo's observations of the phases of Venus falsify the Ptolemaic theory. This is a conditional falsification: if the telescope is reliable (auxiliary hypothesis) and Venus is almost "full" when far away from the earth (test statement), then the

Ptolemaic theory is false. This falsification depends not only on a test statement, but also on an auxiliary hypothesis (cf. appendix below, theorem 3 and the text following it). Galileo assumed that the auxiliary hypothesis is true, that the telescope is reliable, and concluded that the Ptolemaic theory is false. This does not show, as Feyerabend thinks, that Galileo's argumentation was propagandistic and rhetorical. Galileo's auxiliary hypothesis on the reliability of the telescope was intersubjectively testable and critically discussed. Other astronomers accepted Galileo's observations with the telescope because they could repeat them, not because Galileo argued rhetorically or ingeniously. Not even the most ingenious rhetoric would have been sufficient to convince the other astronomers of the phases of Venus and the moons of Jupiter, if these phenomena could not be observed in the telescope by others.

Does reproducibility show that a phenomenon is real? Many illusions, for example the irradiation of bright stars discussed above, are reproducible. However, the problem is not to prove that phenomena are real or that observations are reliable. The problem is to criticise test statements and auxiliary hypotheses, to criticise the claim that a phenomenon is real. In order to show that an observation is illusory, you have to present some reproducible counterexperiment, as Galileo did when he argued that the irradiation of bright stars is illusory. In his discussion of the reliability of the telescope, Feyerabend does not discuss any reproducible counterexperiment, only isolated observations, such as Galileo's drawings of the moon or some vain attempts to see the moons of Jupiter. Therefore Galileo's arguments against the reliability of the naked eye have much more force than Feyerabend's arguments against the reliability of the telescope.

2. *Galileo's New Theory of Motion*

The most important way to test the Ptolemaic and Copernican theories was to observe the positions of the planets. In this way the whole theories were tested, not their constituent hypotheses. But there were other tests. The tower experiment discussed by Ptolemy and Galileo (above, chap. 5, C 3.2) was interesting because it allowed an independent test of a central hypothesis in the Copernican theory, the hypothesis of the diurnal motion of the earth. In his theory Ptolemy assumed that the earth does not move. Already in antiquity heliocentric theories with a moving earth were suggested, but they were not accepted by the ancient astronomers. Ptolemy argued that simple physical experiments refute the hypothesis of a moving earth. Galileo discussed this problem in the following way. If, as Aristotle maintains, every motion presupposes a moving force and if the earth moves (diurnally around its axis), then a stone falling from a tower does not fall vertically down, but will hit the ground far away from the base of the tower. (During the fall there is no force moving the stone in the direction of the assumed rotation of the earth.) However, stones do fall vertically down from towers. Hence the hypothesis of the (diurnal) motion of the earth is falsified. This falsification presupposes the Aristotelian theory of motion as an auxiliary hypothesis: the hypothesis of the (diurnal) motion of the earth is tested together with the Aristotelian theory of motion. However, since Ptolemy and other ancient astronomers regarded the Aristotelian theory of motion not only as unproblematic, but even as true, they thought that the tower experiment falsified the hypothesis of the motion of the earth. If the Aristotelian theory of motion is assumed to be unproblematic, then the hypothesis of the motion of the earth is indeed falsified by the tower experiment. Since Aristotelian physics was not questioned, this refutation was regarded as conclusive for many centuries. This explains why heliocentric theories were not taken seriously for such a long time.

With the Galilean theory of motion the tower experiment no longer falsifies the hypothesis of the motion of the earth. His theory of inertia has the consequence that only changes of motion require a moving force. Due to inertia a stone falls vertically down from a tower independent of whether or not the earth moves. With Galileo's new theory of motion the tower experiment is no longer crucial for the hypothesis of the motion of the earth. Due to inertia the stone follows the motion of the earth during the fall. Thus Galileo's theory eliminates an earlier refutation of the heliocentric theory. Does this show that Galileo immunised the Copernican theory against criticism, as Feyerabend (above, chap. 5, C 3.2) thinks?

At the time of Galileo and Copernicus the Aristotelian theory of motion was regarded as problematic. Only scientists still keeping to the Aristotelian theory, as for example Tycho Brahe, thought that the hypothesis of the motion of the earth was falsified by the tower experiment or by similar experiments. Philosophers in Paris criticised Aristotle's theory during the Middle Ages and suggested a new theory of motion, namely the impetus theory. This theory has an important consequence for the tower experiment: if the earth moves, a stone falls vertically from a tower. Due to an inward force (an impetus) the stone follows the assumed rotation of the earth during its fall (Andersson 1991b; Dijksterhuis 1961, 216–19; Kuhn 1957, 114–22). With the impetus theory as an auxiliary hypothesis the tower experiment no longer falsifies the hypothesis of the motion of the earth. Therefore the impetus theory was important for the development of astronomy. It makes it possible to suggest heliocentric theories. At the end of the fourteenth century the impetus theory had replaced the Aristotelian dynamics in the most important European universities. Copernicus learned the theory in Padua, Galileo in Pisa (Kuhn 1957, 119). While Ptolemy regarded the heliocentric theory as falsified through the tower experiment, the situation was quite different after the impetus theory had been suggested. With this theory as an auxiliary hypothesis the tower experiment agrees with the heliocentric theory and the hypothesis of the motion of the earth. Since the impetus theory was widely spread during Copernicus' and Galileo's time, there was no need to adjust the heliocentric theory to the tower experiment. For this reason Galileo did not have to introduce any new auxiliary hypotheses or to develop a new theory of motion. Galileo did not introduce his new physics in order to help the Copernican theory or in order to immunise it from criticism, but to advance the science of physics (Dijksterhuis 1961, 333–59).

Even if Galileo had suggested his new theory of motion in order to make the Copernican theory compatible with the tower experiment, this would be unimportant for a methodological evaluation of the new theory. According to Popper it is unimportant *when* and *why* new theories are suggested. Feyerabend uses '*ad hoc* hypothesis' in another sense than Popper. Feyerabend regards Galileo's theory of motion as *ad hoc* because he thinks that Galileo suggested it to save the Copernican theory from criticism.

Popper requires that new auxiliary hypotheses should increase the testability (falsifiability) of a theory. As was argued above (chap. 7, B) all auxiliary hypotheses introduced into a system meet this requirement. The testability of the Copernican theory is heightened by the introduction of the telescope and the auxiliary hypothesis that it is reliable. With the telescope many new observations were possible. Also the introduction of Galileo's new theory of motion increased the testability of the Copernican theory and made it possible to derive many new test statements that could not be derived from the Copernican theory alone. According to Popper's definition Galileo's new theory of motion or his auxiliary hypothesis about the telescope was no *ad hoc* hypothesis.

Feyerabend has argued that Aristotelian physics was much more general than Galilean and hence more falsifiable (Feyerabend 1981b, 328). However, this is irrelevant in order to judge if the Galilean theory of motion is *ad hoc* or not. For such judgements we do not compare the Aristotelian and Galilean theories of motion. Of relevance is only if the introduction of these theories as auxiliary hypotheses increases the testability of the tested theory. As argued above, the introduction of the Galilean theory of motion and the introduction of the telescope increased the testability of the Copernican theory.

Popper also suggests that the introduction of new auxiliary hypotheses should be regarded as an attempt to introduce new theoretical systems. According to this idea, the introduction of Galileo's theory of motion led to the discussion of a new theoretical systems, of the Copernican theory together with Galileo's theory of motion. In the same way, the introduction of the telescope also led to the discussion of a new theoretical system, the Copernican theory together with the auxiliary hypothesis that the telescope is reliable. These theoretical systems have to be discussed, tested, and compared with other theoretical system, for example with the Ptolemaic theory together with appropriate auxiliary hypotheses about optics and dynamics. Only empirical testing can decide whether these systems are satisfactory. At least you cannot argue, as Feyerabend does, that the introduction of new auxiliary hypotheses is *ad hoc* or not in agreement with a criticist conception of science. On the contrary, the introduction of new auxiliary hypotheses modifies the discussed theoretical systems and thus shows that criticism is taken seriously.

CHAPTER EIGHT

CRITICAL RATIONALISM AND THE HISTORY OF SCIENCE

A. THE HISTORY OF SCIENCE AND RELATIVISM

1. *Kuhn's and Feyerabend's Historical Relativism*

Thomas Kuhn and Paul Feyerabend criticise the falsificationist conception of science. One of their most important arguments is that the test statements used in science are fallible and impregnated with theories. According to Kuhn, Popper has realised that test statements are fallible without drawing the necessary methodological consequences. Kuhn asks if falsificiations are possible with fallible test statements and how fallible tests statements are chosen by scientists. Kuhn thinks that Popper has no answer to these questions and hence that the falsificationist conception of science is unrealistic. This is obvious when scientific revolutions occur, he thinks. Then the basic theory is changed. As a result the world is seen in a new way, and new test statements are accepted. Kuhn compares this change to a gestalt switch and even says that scientists live in another world after a scientific revolution. For example, before the Copernican revolution astronomers lived in a world in which the moon is a planet, and afterwards in a world in which the moon is a satellite. Therefore there are no neutral test statements that can be used to test competing theories, and scientific theories cannot be rationally compared, but are incommensurable. It also follows that scientific revolutions are not the result of objective empirical tests, but rather of propaganda, persuasion, and conversions (chapter 3, A).

Paul Feyerabend argues in the same way. He maintains that theories are partially incommensurable. Although the Ptolemaic and Copernican theories have common test statements about the positions of the planets and are commensurable in this respect, the two theories were optically and dynamically incommensurable, as Galileo's defence of the Copernican theory shows. Galileo used the telescope to support the Copernican theory. However, his observations were not accepted by those defending the Ptolemaic theory, who only accepted observations with the naked eye (chapter 5, C 1.1 above).

Those defending the Ptolemaic theory also maintained that simple physical experiments falsified the Copernican theory, as for example the tower experiment. According to Feyerabend, Galileo interpreted this experiment in a new way to avoid a falsification of the Copernican theory. This new "natural interpretation" of the tower experiment shows that the Ptolemaic and Copernican theories are dynamically incommensurable. In this way most competing scientific theories are at least partially incommensurable. A rational discussion of competing theories has only limited value. Kuhn and Feyerabend think that propaganda, persuasion, and conversions are more important for a change of theories to take place (chapter 5, C 1.2, above).

Kuhn argues that in the history of science falsifications are seldom taken seriously, but are regarded as anomalies to be digested by the ruling paradigm. Since falsifications seldom lead to a total rejection of theories or to scientific revolutions, Kuhn thinks that critical rationalists have unrealistic ideas about the importance of falsifications in science (above, chapter 3, B).

Feyerabend argues in a similar way: in the history of science, falsified theories can often be modified and developed. This shows that it is unreasonable to reject them totally. New theories often meet with considerable difficulties before the necessary auxiliary hypotheses have been developed, as happened in the Copernican revolution. Only after many years were the optical and dynamic theories necessary for the Copernican revolution developed and able to replace the old Aristotelian conception of the universe (above, chapter 5, C 1).

2. *Lakatos's Attempt to Save Reason in Science*

Lakatos discusses different ideas about test statements, among others the positivist idea that test statements can be proved by experience. Lakatos is aware of the fact that Popper rejects this conception of test statements as dogmatic (above, chapter 4 A).

What is Popper's conception of test statements if he rejects the positivist idea of infallible test statements? Lakatos thinks that Popper recommends that test statements should be chosen by conventional decisions that make test statements (and falsificiations) infallible (above, chapter 4, B).

Lakatos does not find this conventionalist solution to the problem satisfactory and asks how test statements can be criticised. For a critical conception of science this is an important question. Lakatos thinks that it is too arbitrary to select test statements merely by conventional decisions. He does not think that test statements can be criticised by their deductive consequences. Rather Lakatos thinks that test statements are criticised indirectly by criticism of the "interpretative theories" used to interpret experience. If Lakatos had been consistent, this idea would have led him to the problem of incommensurability: we can use different interpretative theories when we discuss different theories. Lakatos does not discuss this problem and has no solution to the problem of how a rational discussion of fallible test statements is possible. In order to save reason in science, a satisfactory solution to this problem is necessary (above, chapter 4, B).

Lakatos also discusses falsificiations that do not lead to the total rejection of theories. He thinks that this problem shows that falsificationism is unrealistic. In Lakatos's own methodology falsificiations (in the logical sense) have no significance; theories and research programmes are rejected according to new rules of elimination. Kuhn thinks that a paradigm is rejected when there is a new paradigm that seems to be better. Lakatos adopts this idea, but, differing from Kuhn, he thinks that it is not merely a question of subjective belief to judge which paradigm, theory, or research programme is better. According to Lakatos, objective and rational criteria can be given for this choice. This is the kernel of Lakatos's attempt to save reason in science. However, when Lakatos develops his ideas concerning the choice of theories and research programmes, they have strange consequences. First Lakatos tries to tell scientists which theories and research programmes they should work on. However, the criticisms of Feyerabend and others make him realise some problems in this approach, and as a result he withdrew from methodology to rational reconstructions of the history of science (above, chapter 4, E). This is a catastrophic retreat, which shows that Lakatos's attempt to save reason in science fails. Lakatos's attempts to reconstruct the history of science rationally are highly problematic, too.

Lakatos tried to answer the problems raised by the study of the history of science. Since he does not have any satisfactory solutions to this problem, it seems as if the criticism of falsificationism is well grounded. In order to discuss this problem I began with a discussion of the logical structure of falsifications.

3. *Logical Structure of Falsifications*

Popper discusses two types of falsifications, falsifications of theoretical systems and of isolated hypotheses. When Popper discusses the first case, falsifications of theoretical systems, he only discusses a special case: falsification of a theoretical system consisting of general hypotheses and initial conditions. In the second case he discusses the falsification of an isolated general hypothesis. In this case he is confronted with another limitation: in science we often test theories consisting of many hypotheses.

In the chapter two I showed that there are general relations between the derivations of predictions and falsifications. If an empirical prognosis can be derived from general hypotheses and singular initial conditions, a number of different types of falsifications are valid. General (meta)logical arguments show this. If a deductive argument is valid, then the conclusion must be true when the premises are true. Thus, if the conclusion is false, all premises cannot be true. Therefore, if a prediction is false and derived from general hypotheses and singular initial conditions, at least one of the premises must be false, that is, at least one hypothesis or initial condition must be false. This is the first type of falsifying argument dealt with by Popper.

In many cases the initial conditions describe observable events. If this is the case, we can maintain that observations show that some or all initial conditions are true, and these initial conditions can be used as falsifying premises. Then we get another type of falsification in which at least some of the initial conditions are used as falsifying premises. If all the initial conditions describe observable events, we can test and perhaps falsify a theoretical system consisting only of universal hypotheses (above, chapter 2, C).

Popper deals with a special case of this second type of falsification, the case in which only an isolated universal hypothesis is tested. He argues that such hypotheses can be tested and refuted by "basic statements." For example, the general hypothesis 'all swans are white' is refuted by the basic statement 'this swan is not white'. The metalogical considerations discussed above allow a much more general solution to the problem of which hypotheses and theoretical systems are falsifiable. A theory is falsifiable, if (and only if) it has empirical consequences. Then the theoretical system consisting of the conjunction of the general hypotheses and initial conditions is falsifiable. If the initial conditions are empirical and describe observable events, then we can test a smaller theoretical system consisting only of the conjunction of the general hypotheses.

In many cases a part of the theoretical system is testable (and falsifiable). This is the case when a part of the theoretical system is sufficient to derive an empirical prognosis. Then this part can be tested in isolation. In some cases even an isolated general hypothesis can be tested in this way. Thus many different types of theoretical systems can be tested.

In order to derive empirical consequences, we often need auxiliary hypotheses. These hypotheses then become parts of the tested theoretical system. However, it is possible that auxiliary hypotheses can be tested independently in other empirical tests. Such an independently testable auxiliary hypothesis is for example Galileo's hypothesis that the telescope is reliable. When an auxiliary hypothesis is testable in this way and is corroborated, then it can be regarded as a part of the unproblematic background knowledge. In this case a third type of falsification is interesting: falsificiations in which auxiliary hypotheses are regarded as parts of the falsifying premises (text after theorem 3 in the appendix below).

B. Experience Impregnated by Theory

1. *Falsification with Fallible Test Statements*

Kuhn doubts that falsifications are possible with fallible test statements. This doubt is only justified if we presuppose that falsificiations must be infallible. But there is no reason for presupposing this. From a logical point of view, a theory is false if some test statements (as falsifying premises) are true. According to a fallibilist point of view, falsificiations are just as fallible as the falsifying test statements. Test statements and falsifications can be criticised. If the test statements used in a falsification are false, then the falsification must have been wrong and has to be withdrawn. In the history of science there are a number of such withdrawn falsifications. This does not make falsifications uninteresting, in the same way as the fallibility of general hypotheses does not make hypotheses uninteresting. On the contrary, from a methodological and historical point of view the fallibility of falsifications is very interesting.

2. *Criticism of Test Statements*

Can test statements be criticised? One possible type of criticism is to repeat observations and experiments. If such repetitions give other results, then the test statements are problematic. When, for example, Galileo declared that Jupiter is surrounded by moons, other astronomers tried to reproduce his observations. In the beginning this was not so easy (above, chapters 5 C 1.1, and 6 D 2). However, after a while the astronomers agreed that the moons of Jupiter exist. They had tested Galileo's observations and had obtained positive results. In the same way astronomers tried to repeat Herschel's observations of Uranus and Galle's observations of Neptune. Since other astronomers could also repeat these observations, they were soon generally accepted.

Such tests of test statements presuppose some background knowledge as auxiliary hypotheses (*AH*). There must be some regularities telling the scientists how to reproduce the problematic observations and experiments. In this way we can test a test statement *Pk* by deriving a test statement *Pl* of the same type with the help of an auxiliary hypothesis (*AH*):

(1) $Pk, AH \vdash Pl$

Leverrier assumed that there is an unknown planet, which he called Vulcan, between Mercury and the sun. At first some astronomers reported that they had observed this planet. However, these observations could never be reproduced in a regular way. Astronomers trying to observe the hypothetical planet in the predicted positions did not see anything. These negative results were reproducible and were used to criticise the first positive reports. In this way the test statements on the planet Vulcan were criticised and rejected.

We can falsify a test statement if we can show that one of its deductive consequences is false. Assume that in (1) *Pk* is a test statement about Vulcan (in the space-time region *k*) and that with some unproblematic auxiliary hypothesis about the movement of planets (*AH*) we derive another test statement *Pl* about Vulcan. Assume that the observations of Vulcan cannot be repeated, that is that *Pl* is false, and that we do not question our background knowledge *AH*. Then we can falsify *Pk* (see below, appendix, theorem 3(d), $r=m=n=1$, $k=0$, $C_1=Pk$, $H_1=AH$, $P=Pl$):

(2) $\neg Pl, AH \vdash \neg Pk$

This shows how test statements can be tested by the derivation of other test statements of the same type, by the derivation of homotypic test statements (above, chapter 6 C 4). In science as in ordinary life, our first reaction to a problematic observation is to try to repeat it, to look once again.

It is also possible to derive test statements of a new type, that is to derive heterotypic test statements. This type of criticism has not only to do with the reproducibility of observations and experiments, but is deeper. For example, we can doubt that a test statement describes an observable event. In other cases, scientists report differently about the same observation or experiment. When Herschel first observed Uranus, he thought that he had discovered a new comet. Earlier astronomers observing the same object thought that they saw a fixed star. After some months Herschel changed his opinion and thought that he had discovered a new planet. Here we have three types of different test statements (about a fixed star, a comet, and a planet), which in the beginning were problematic. They could be tested and critically discussed. The background knowledge told Herschel that a fixed star ought to look like a luminous point when observed in a telescope. From a test statement about a fixed star (*Pl*) we can derive a test statement about a luminous point (*Qk*) with the help of background knowledge (*AH*):

(3) $Pl, AH \vdash Qk$

In this way a test statement of a new type, a heterotypic test statement, is derived. Since Herschel saw that the problematic object looked like a luminous disk, he could falsify the assumption that it was a fixed star with the help of background knowledge (see below, appendix, theorem 3(d), $r=m=n=1$, $k=0$, $C_1=Pk$, $H_1=AH$, $P=Ql$):

(4) $\neg Ql, AH \vdash \neg Pk$

In a similar way he could later falsify the assumption that he had observed a comet, since the object did not move like a comet.

In the history of science we can find many examples of such criticism of test statements. They illustrate how test statements can be criticised with their deductive consequences (above, chapter 6, A 2 and D).

Is it possible to criticise test statements with interpretative theories, as Lakatos suggests? The derivations of homo- and heterotypical test statements above show that auxiliary hypotheses are important in order to criticise test statements. They also show that test statements are not accepted as true on the basis of auxiliary hypotheses (or interpretative theories), but are rather criticised with their help. Herschel, for example, rather than criticise an auxiliary hypothesis used it to reject the assumption that Uranus is a fixed star. Therefore, the use of auxiliary hypotheses in the discussion of test statements does not show that test statements cannot be criticised with their deductive consequences, but rather that auxiliary hypotheses are used to derive such consequences.

3. *Incommensurability of Competing Theories?*

Kuhn and Feyerabend think that a rational discussion of competing theories is impossible because such theories are incommensurable. They think that scientists with different theoretical presuppositions not only interpret experience in different ways, but even see the world differently and perhaps even live in different worlds. In such cases scientists with different theoretical presuppositions accept different test statements. For example, scientists accepting the Ptolemaic theory see the world in a Ptolemaic way, and scientists accepting the Copernican theory see the world in a Copernican way. A rational discussion is no longer possible, and scientists have to use irrational means, such as propaganda and persuasion.

Let us assume that competing theories are incommensurable in the sense that different types of test statements are accepted. Then there are no generally accepted unproblematic test statements. From a criticist point of view this does not show that the competing theories are necessarily incommensurable, but rather that there is a problem. Above I showed how from one type of test statement another type of test statement can be derived. Can the derivation of heterotypic test statements be used to solve the problem of incommensurability? If one type of test statement is problematic, is it possible to derive another and unproblematic type of test statement?

From a logical point of view the problem of incommensurability can be stated in the following way. Assume that scientists with different fundamental theories see different things in the same experiment, for example phlogiston (*Pk*) and oxygen (*Qk*). Assume further that they maintain that these experiments are reproducible, that scientists believing in the phlogiston theory always see phlogiston in the same experiments in which scientists believing in the oxygen theory see oxygen. Does this show that a rational discussion no longer is possible, that it is time to begin with propaganda and persuasion in order to convert those believing in the competing theory?

From a criticist point of view, there is a problem: the problem of which test statements are true. Is there phlogiston or oxygen in the experiment? This can be discussed in the ordinary way by deriving consequences from problematic statements. Since there is no agreement on the test statements about the experiment, these test statements are problematic. Assume that the oxygen theory (*AH*) tells us that in oxygen a candle will shine brightly. Then we have succeeded in deriving unproblematic test statements from problematic ones:

(5) $Pk, AH \vdash Rk$

This derivation tells us that if there is oxygen in a space-time region, then a candle will shine brightly in the same region. Assume further that the phlogiston theory (*AH´*) tells us that if the air is full with phlogiston in a space-time region (*Qk*), then a candle will go out in this region:

(6) $Qk, AH' \vdash \neg Rk$

In this case the problem of incommensurability has been solved by the derivation of common and unproblematic test statements. Scientists may disagree on whether there is oxygen or air full with phlogiston in a certain time-space region, but they can agree on whether or not a candle shines brightly.

It is not certain that the unproblematic consequences are different, as in the example above. If *Sk* stands for the statement that the air is without phlogiston in a space-time region, that is, that the air is "dephlogisticated," then the unproblematic consequence follows that a candle will shine brightly (*Rk*):

(7) $Sk, AH' \vdash Rk$

In this case both competing theories have the same consequence, that a candle will shine brightly.

Thus the fact that test statements are impregnated with theories does not necessarily have the consequence that competing theories are incommensurable. If both theories allow the derivation of unproblematic test statements, then the problem of incommensurability is solved. Is it always possible to derive unproblematic test statements from competing theories? In the preceding chapters I tested the best examples discussed by Kuhn and Feyerabend and showed that in all these examples unproblematic test statements could be derived.

The phlogiston and oxygen theories were perfectly commensurable, in spite of Kuhn's assertion that Priestley "saw" dephlogisticated air where Lavoisier "saw" oxygen. In a similar way, the Ptolemaic and Copernican theories were commensurable, in spite of Feyerabend's correct assertion that an astronomer saw the planets differently with the telescope than with the naked eye. Thus the two theories were not optically incommensurable. Feyerabend's assertion that the two theories were dynamically incommensurable is also false. Both Copernicus and Ptolemy accepted the same test statements about the tower experiment: a stone falling freely from a tower falls vertically down. Due to different auxiliary hypotheses they derived different conclusions. This has nothing to do with different test statements, but with different auxiliary hypotheses.

The impression that competing theories are incommensurable arises if fallible and theory-impregnated test statements cannot be rationally discussed. For a positivist philosophy of science assuming absolute and infallible test statements, the fallibility and theory-impregnation of test statements is a serious problem, but for a criticist conception of science, this is not the case. On the contrary, in the discussion of such test statements the criticist conception of science shows its strength. A critical discussion of problematic test statements is possible. As my analysis above and in the preceding chapters shows, such a critical discussion of test statements is usual in the history of science. The idea that competing theories are incommensurable is inspired not so much by the history of science as by a mistaken philosophy of science. What Popper calls the myth of the framework has its origin in philosophy and is the basis of Kuhn's and Feyerabend's belief that competing theories cannot be rationally discussed, that scientists have to use propaganda and persuasion to convert other scientists to new theories. According to Popper (1970b, 56) the myth of the framework is the central bulwark of modern irrationalism. This myth is the source of Kuhn's and Feyerabend's belief that competing theories are incommensurable.

C. Immunity Against Criticism

Kuhn and Feyerabend maintain that there are always anomalies and hence that most theories are falsified, but that there are many ways to immunise theories against falsificiations. They think that such immunisation of theories is important and useful for the development of science. Lakatos was so impressed by these ideas that he thought that falsifications hardly play any role in the development of science and advanced his methodology of scientific research programmes, in which positive cases supporting a theory or research programme are important, not negative and falsifying. Kuhn, Lakatos, and Feyerabend discuss the following ways to immunise falsifications: (1) to modify a falsified theory, (2) to introduce *ad hoc* hypotheses or (3) to introduce auxiliary hypotheses.

1. *Theoretical Modifications*

In order to discuss the first way to avoid falsifications, to modify falsified theories, a methodological discussion of falsifications is necessary, rather than a discussion of cases from the history of science. Without doubt, falsified theories are often more or less extensively modified. A falsification shows that at least some part of the theory is false, and hence that the conjunction of the hypotheses of the theory is false. But it does not show which parts of the theory are false, whether central or peripheral hypotheses have to be changed. It is dogmatic to maintain that falsified theories always have to be rejected completely. We can only guess how a falsified theory is to be changed, whether it is to be completely rejected or only partially modified. It is not the task of methodology to state how to change falsified theories. We can only suggest different changes and test them empirically. There is no method of discovery. This is why Lakatos's methodology of scientific research programmes fails. If a falsified theory is modified, this shows that the falsification is taken seriously, and hence that an attempt is made to learn from experience. Much of Kuhn's, Lakatos's, and Feyerabend's criticism of falsificationism rests on the mistaken idea that falsified theories have to be completely rejected.

2. Ad Hoc *Hypotheses*

Is it always possible to avoid falsifications by introduction of *ad hoc* hypotheses? It is difficult to discuss this question because it is unclear which hypotheses are *ad hoc*.

Popper discusses three different definitions of an *ad hoc* hypothesis: (1) a hypothesis introduced in a theory that reduces the testability (falsifiability) of a theory, (2) a hypothesis introduced in a theory that is not independently testable, and (3) a hypothesis that is introduced to criticise test statements (above chapter 7 C and D 1).

The first definition is too narrow. It is impossible to reduce the testability of a theory by introducing additional hypotheses into it. For the same reason, it is also impossible to avoid falsifications in this way. If a theory is false, then the theoretical system consisting of the theory and an additional hypothesis is also false. The second definition is too wide. If you demand generally that theoretical modifications have to be independently testable, this would be an obstacle to the development of science. Above (chapter 7 B) I discussed Pauli's introduction of the neutrino-hypothesis. Another example is Kepler's modification of the Copernican theory. This modification did not lead to any new types of predictions, but Kepler tested his theory as Copernicus had done, by observing the positions of the planets. Although Kepler's theory was not independently testable, it was an important step forward.

According to the third definition, an *ad hoc* hypothesis is not introduced into a theoretical system, but is used to criticise test statements. Such criticism may be raised for good reasons or simply in order to immunise theories against criticism. If test statements are criticised without any empirical or theoretical reasons in order to avoid falsifications, then the hypotheses introduced for this reason can be rejected as *ad hoc*. However, if there are empirical or theoretical reasons to question test statements, the criticism has to be taken seriously and cannot be rejected as *ad hoc*. Since test statements are fallible, it is in principle possible that they are mistaken. In such cases scientists have to try to derive unproblematic test statements about intersubjectively testable effects in order to decide what test statements can be used in the critical discussion of theories. It also possible that implicit auxiliary hypotheses used in order to interpret experience are mistaken. Then scientists have to discover which auxiliary hypotheses are problematic and change them.

3. *Auxiliary Hypotheses*

Feyerabend thinks that Galileo immunised the Copernican theory from criticism by introducing new auxiliary hypotheses. These hypotheses were introduced into the theoretical systems and thus were not *ad hoc* in the sense that they questioned isolated test statements only. Genuine *ad hoc* hypotheses are not parts of the tested theory, but are only used to question some of the test statements. If the hypotheses are introduced into the theoretical system, if they are used to derive empirical consequences, then the situation is different. We have a new kind of theoretical system. By the introduction of the auxiliary hypotheses, the theoretical system has changed. Therefore, the introduction of the new hypotheses has not immunised a theory against criticism, but has changed it. This is an attempt to learn from experience.

Galileo's new dynamics was an auxiliary hypothesis used to test the Copernican theory. From the hypothesis of the diurnal motion of the earth and Galileo's dynamics it follows that a stone falls vertically from a tower. Thus Galileo's dynamics was auxiliary in this test, rather than *ad hoc*.

Galileo's hypothesis that the telescope is a reliable instrument can be regarded in a similar way. It was used to derive new predictions about the results of telescopic observations. Thus it can be regarded as a part of the tested theoretical system.

Before the introduction of the telescope and Galileo's theory of motion other auxiliary hypotheses were used, namely the hypothesis that observations with the naked eye are reliable and the Aristotelian theory of motion. With these older auxiliary hypotheses the Copernican theory was falsified, Feyerabend argues.

This argument is not quite correct, since the Copernican theory T was falsified together with an auxiliary hypothesis AH. Thus the theoretical system (T & AH), rather than only T, was falsified. By such falsifications we do not know whether the theory T or the auxiliary hypothesis AH is false. Assume that a new auxiliary hypothesis AH' is introduced and that the theoretical system (T & AH') is corroborated. Is this a case of immunisation of a theory against criticism? Is the theory T saved by introducing *ad hoc* the new auxiliary hypothesis AH'? No, the theoretical system (T & AH) has been changed into (T & AH'): The result of the critical discussion is not immunisation from criticism, but theoretical change. Whether the theoretical change is acceptable can only be decided by empirical tests of the new theoretical system, not by abstract methodological considerations.

In some cases new auxiliary hypotheses can be tested in isolation and not only as parts of theoretical systems. The auxiliary hypotheses discussed above, Galileo's hypothesis that the telescope is reliable and his new dynamics, could be tested in isolation. In such cases the auxiliary hypotheses should of course be tested in isolation. If the auxiliary hypotheses are corroborated in such tests, they need not be regarded as parts of the tested system, but can be used as unproblematic background knowledge and as falsifying premises. For example, when the reliability of the telescope has been tested and corroborated, we can argue that the Ptolemaic theory is falsified. In this falsification, the auxiliary hypothesis is used as a falsifying premise (see below, appendix, theorem 3(d) and comment):

(8) $C, AH, \neg P \vdash \neg T$

Thus auxiliary hypotheses can be parts of the tested theoretical system or parts of the statements used to test it. How they are treated depends on whether they are independently testable.

Auxiliary hypotheses used to derive empirical consequences are not used to immunise theories against criticism; on the contrary, they are used to test theories severely. The impression of immunisation against criticism arises when a theory remains unchanged while auxiliary hypotheses are changed. However, this impression disappears if you take into account that a whole theoretical system consisting of a theory and auxiliary hypotheses is tested and that this system is changed. If this is taken in to account, the use and change of auxiliary hypotheses in no way shows that falsificationism is untenable. On the contrary, auxiliary hypotheses can be used to improve the critical discussion of theories.

D. The History of Science and Criticism

Kuhn and Feyerabend criticise the philosophy of science of Critical Rationalism with the help of the history of science. Impressed by this criticism Lakatos developed his methodology of scientific research programmes to avoid relativism. This attempt to save a rationalist conception of science is highly problematic and ends in a rational reconstruction of the history of science. From a criticist point of view Lakatos's attempt to understand the historical development of science is unsatisfactory.

Kuhn's and Feyerabend's criticism is based on two important methodological problems: the problem that experience is fallible and impregnated with theory, and the problem of theoretical modifications. The first problem leads them to think that competing theories are incommensurable. Above I developed a criticist conception of experience and of test statements according to which a critical discussion of test statements is possible, although test statements are fallible and impregnated with theories. Therefore scientists do not have to use propaganda and persuasion in their discussions. I also argued that the problem of incommensurability can be solved in a criticist way: unproblematic (and "commensurable") test statements can be derived from problematic (and "incommensurable") ones.

The history of science shows that falsified theories are seldom totally rejected, but that scientists often try to modify them more or less extensively. These modifications are not attempts to immunise criticism, but to learn from experience. The introduction and changing of auxiliary hypotheses can be judged in a similar way. Such changes are also attempts to learn from experience. Auxiliary hypotheses do not have to remain implicit, but can be explicitly stated and critically discussed, either in isolation or as parts of theoretical systems.

The two basic problems in this discussion can be solved in a criticist way. This criticist conception allows a deeper understanding of the history of science and contributes more to the growth of knowledge than the relativist conceptions suggested by Kuhn and Feyerabend.

APPENDIX

THEOREMS ON EXPLANATION, PREDICTION, AND FALSIFICATION

The following metalogical theorems are proved with the rules for introduction and elimination of connectives, that is, with the rules for natural deductions (Grize 1972, 24–25; Kleene 1967, §§ 9–13). As usual, the metalogic is expressed in natural language and not formalised, except the sign for deducibility ($\vdash$) (cf. Grize 1972, 3–6, 36–37).

THEOREM 1:

(a) $H, C \vdash P$, iff $H \vdash C \rightarrow P$

(b) $H_1, \ldots, H_r, C_1, \ldots, C_n \vdash P$, iff
$H_1, \ldots, H_r, C_1, \ldots, C_k \vdash (C_{k+1} \wedge \ldots \wedge C_n) \rightarrow P$ $(0 \leq k \leq n)$

PROOF:

(a)

I: Presupposition: $H, C \vdash P$ (deduction in k steps).

1	H	
2	$\quad C$	
⋮	$\quad \vdots$	
k+2	$\quad P$	1, 2, presupposition
k+3	$C \rightarrow P$	2 – (k+2), $\rightarrow$ I

Thus: If $H, C \vdash P$, then $H \vdash C \rightarrow P$.

II: Presupposition: $H \vdash C \rightarrow P$ (deduction in m steps).

1	H	
2	C	
⋮	$\vdots$	
m+2	$C \rightarrow P$	1, presupposition
m+3	P	2, m+2, $\rightarrow$ E

Thus: If $H \vdash C \rightarrow P$, then $H, C \vdash P$.
Theorem 1 (a) follows from I and II.

(b)

I: Presupposition: $H_1, \ldots, H_r, C_1, \ldots, C_n \vdash P$ (deduction in p steps).

1	H_1	
$\vdots$	$\vdots$	
r	H_r	
$r+1$	C_1	
$\vdots$	$\vdots$	
$r+k$	C_k	
$r+k+1$	$\quad C_{k+1} \wedge \ldots \wedge C_n$	
$r+k+2$	$\quad C_{k+1}$	$r+k+1, \wedge$ E
$\vdots$	$\quad \vdots$	
$r+n+1$	$\quad C_n$	$r+k+1, \wedge$ E
$\vdots$	$\quad \vdots$	
$r+n+p+1$	$\quad P$	$1, \ldots, r+k, r+k+2, \ldots, r+n+1$, presupposition
$r+n+p+2$	$(C_{k+1} \wedge \ldots \wedge C_n) \rightarrow P$	$(r+k+1) - (r+k+p+1), \rightarrow$ I

Thus: If $H_1, \ldots, H_r, C_1, \ldots, C_n \vdash P$, then $H_1, \ldots, H_r, C_1, \ldots, C_k \vdash (C_{k+1} \wedge \ldots \wedge C_n) \rightarrow P$.

II. Presupposition: $H_1, \ldots, H_r, C_1, \ldots, C_k \vdash (C_{k+1} \wedge \ldots \wedge C_n) \rightarrow P$ (deduction in q steps).

1	H_1	
$\vdots$	$\vdots$	
r	H_r	
$r+1$	C_1	
$\vdots$	$\vdots$	
$r+k$	C_k	
$r+k+1$	C_{k+1}	
$\vdots$	$\vdots$	
$r+n$	C_n	
$\vdots$	$\vdots$	
$r+n+q$	$(C_{k+1} \wedge \ldots \wedge C_n) \rightarrow P$	$1, \ldots, r+k$, presupposition
$r+n+q+1$	$C_{k+1} \wedge \ldots \wedge C_n$	$r+k+1, \ldots, r+n, \wedge$ I
$r+n+q+2$	P	$r+n+q, r+n+q+1, \rightarrow$ E

Thus: If $H_1, \ldots, H_r, C_1, \ldots, C_k \vdash (C_{k+1} \wedge \ldots \wedge C_n) \rightarrow P$, then $H_1, \ldots, H_r, C_1, \ldots, C_n \vdash P$. Theorem 1 (b) follows from I and II.

THEOREM 2:

(a) $H, C \vdash P$, iff $\neg P \vdash \neg(H \wedge C)$

(b) $H_1, \ldots, H_r, C_1, \ldots, C_n \vdash P$, iff $\neg P \vdash \neg(H_1 \wedge \ldots \wedge H_r \wedge C_1 \wedge \ldots \wedge C_n)$

PROOF:

(a)

I. Presupposition: $H, C \vdash P$ (deduction in m steps).

1	$\neg P$	
2	$\quad H \wedge C$	
3	$\quad H$	2, $\wedge$ E
4	$\quad C$	2, $\wedge$ E
⋮	$\quad$ ⋮	
m+4	$\quad P$	3, 4, presupposition
m+5	$\quad \neg P$	1, repetition
m+6	$\neg(H \wedge C)$	2, m+4, m+5, $\neg$ I

Thus: If $H, C \vdash P$, then $\neg P \vdash \neg(H \wedge C)$.

II. Presupposition: $\neg P \vdash \neg(H \wedge C)$ (deduction in n steps).

1	H	
2	C	
3	$\quad \neg P$	
⋮	$\quad$ ⋮	
n+3	$\quad \neg(H \wedge C)$	3, presupposition
n+4	$\quad H \wedge C$	1, 2, $\wedge$ I
n+5	$\neg\neg P$	3, n+3, n+4, $\neg$ I
n+6	P	$\neg$ E

Thus: If $\neg P \vdash \neg(H \wedge C)$, then $H, C \vdash P$.
Theorem 2(a) follows from I and II.

(b)
I. Presupposition: $H_1, \ldots, H_r, C_1, \ldots, C_n \vdash P$ (deduction in q steps).

1	$\neg P$	
2	$\quad H_1 \wedge \ldots \wedge H_r \wedge C_1 \wedge \ldots \wedge C_n$	
3	$\quad H_1$	2, $\wedge$ E
⋮	$\quad$ ⋮	
r+2	$\quad H_r$	2, $\wedge$ E
r+3	$\quad C_1$	2, $\wedge$ E
⋮	$\quad$ ⋮	
r+n+3	$\quad C_n$	2, $\wedge$ E
⋮	$\quad$ ⋮	
r+n+q+3	$\quad P$	3, …, r+n+3, presupposition
r+n+q+4	$\quad \neg P$	1, repetition
r+n+q+5	$\neg(H_1 \wedge \ldots \wedge H_r \wedge C_1 \wedge \ldots \wedge C_n)$	2, r+n+q+3, r+n+q+4, $\neg$ I

Thus: If $H_1, \ldots, H_r, C_1, \ldots, C_n \vdash P$,
then $\neg P \vdash \neg(H_1 \wedge \ldots \wedge H_r \wedge C_1 \wedge \ldots \wedge C_n)$.

II. Presupposition: $\neg P \vdash \neg(H_1 \wedge \ldots \wedge H_r \wedge C_1 \wedge \ldots \wedge C_n)$ (deduction in p steps).

1	H_1	
$\vdots$	$\vdots$	
r	H_r	
$r+1$	C_1	
$\vdots$	$\vdots$	
$r+n$	C_n	
$r+n+1$	$\quad \neg P$	
$\vdots$	$\quad \vdots$	
$r+n+p+1$	$\quad \neg(H_1 \wedge \ldots \wedge H_r \wedge C_1 \wedge \ldots \wedge C_n)$	$r+n+1$, presupposition
$r+n+p+2$	$\quad H_1 \wedge \ldots \wedge H_r \wedge C_1 \wedge \ldots \wedge C_n$	$1, \ldots, r+n, \wedge$ I
$r+n+p+3$	$\neg\neg P$	$r+n+1$, $r+n+p+1$, $r+n+p+2$, $\neg$ I
$r+n+p+4$	P	$r+n+p+3$, $\neg$ E

Thus: If $\neg P \vdash \neg(H_1 \wedge \ldots \wedge H_r \wedge C_1 \wedge \ldots \wedge C_n)$,
then $H_1, \ldots, H_r, C_1, \ldots, C_n \vdash P$.
Theorem 2(b) follows from I and II.

THEOREM 3:

(a) $H, C \vdash P$, iff $\neg P, C \vdash \neg H$

(b) $H_1, \ldots, H_r, C_1, \ldots, C_n \vdash P$, iff $\neg P\ C_1, \ldots, C_n \vdash \neg(H_1 \wedge \ldots \wedge H_r)$

(c) $H_1, \ldots, H_r, C_1, \ldots, C_n \vdash P$, iff
$\neg P, C_1, \ldots, C_k \vdash \neg(H_1 \wedge \ldots \wedge H_r \wedge C_{k+1} \wedge \ldots \wedge C_n)$ $(0 \leq k \leq n)$

(d) $H_1, \ldots, H_r, C_1, \ldots, C_n \vdash P$, iff
$\neg P, H_1, \ldots, H_m, C_1, \ldots, C_k \vdash \neg(H_{m+1} \wedge \ldots \wedge H_r \wedge C_{k+1} \wedge \ldots \wedge C_n)$
$(0 \leq m \leq r)$ $(0 \leq k \leq n)$

PROOF:

(a)

I. Presupposition: $H, C \vdash P$ (deduction in p steps).

1	C	
2	$\neg P$	
3	$\quad H$	
$\vdots$	$\quad \vdots$	
$p+3$	$\quad P$	3, 1, presupposition
$p+4$	$\quad \neg P$	2, rep.
$p+5$	$\neg H$	3, $p+3$, $p+4$, $\neg$ I

Thus: If $H, C \vdash P$, then $\neg P, C \vdash \neg H$.

II. Presupposition: If $\neg P, C \vdash \neg H$ (deduction in q steps).

1	H	
2	C	
3	$\neg P$	
⋮	⋮	
q+3	$\neg H$	2, 3, presupposition
q+4	H	1, rep.
q+5	$\neg\neg P$	3, q+3, q+4, ¬ I
q+6	P	q+5, ¬ E

Thus: If $\neg P, C \vdash \neg H$, then $H, C \vdash P$.

Theorem 1(a) follows from I and II.

(b) Theorem 3(b) is a special case of theorem 3(d) (k=n, m=0).

(c) Theorem 3(c) is a special case of theorem 3(d) (m=0).

(d)

I. Presupposition: $H_1, \ldots, H_r, C_1, \ldots, C_n \vdash P$ (deduction in p steps).

1	H_1	
⋮	⋮	
m	H_m	
m+1	C_1	
⋮	⋮	
m+k	C_k	
m+k+1	$\neg P$	
m+k+2	$H_{m+1} \wedge \ldots \wedge H_r \wedge C_{k+1} \wedge \ldots \wedge C_n$	
m+k+3	M_{m+1}	m+k+2, ∧ E
⋮	⋮	
r+k+2	M_r	m+k+2, ∧ E
r+k+3	C_{k+1}	m+k+2, ∧ E
⋮	⋮	
r+n+3	C_n	m+k+2, ∧ E
⋮	⋮	
r+n+p+3	P	1, …, m+k, m+k+3, …, r+n+3, presupposition
r+n+p+4	$\neg P$	m+k+1, rep.
r+n+p+5	$\neg(H_{m+1} \wedge \ldots \wedge H_r \wedge C_{k+1} \wedge \ldots \wedge C_n)$	m+k+2, n+n+p+3, r+n+p+4, ¬ I

Thus: If $H_1, \ldots, H_r, C_1, \ldots, C_n \vdash P$, then $H_1, \ldots, H_m, C_1, \ldots, C_k, \neg P \vdash \neg(H_{m+1} \wedge \ldots \wedge H_r \wedge C_{k+1} \wedge \ldots \wedge C_n)$

II. Presupposition: $H_1, \ldots, H_m, C_1, \ldots, C_k, \neg P \vdash \neg(H_{m+1} \wedge \ldots \wedge H_r \wedge C_{k+1} \wedge \ldots \wedge C_n)$ (deduction in q steps).

1	H_1	
$\vdots$	$\vdots$	
m	H_m	
$\vdots$	$\vdots$	
r	H_r	
$r+1$	C_1	
$\vdots$	$\vdots$	
$r+k$	C_k	
$\vdots$	$\vdots$	
$r+n$	C_n	
$r+n+1$	$\quad \neg P$	
$\vdots$	$\quad \vdots$	
$r+n+q+1$	$\quad \neg(H_{m+1} \wedge \ldots \wedge H_r \wedge C_{k+1} \wedge \ldots \wedge C_n)$	1, …, m, $r+1$, …, $r+k$, $r+n+1$, presupposition
$r+n+q+2$	$\quad H_{m+1} \wedge \ldots \wedge H_r \wedge C_{k+1} \wedge \ldots \wedge C_n$	$m+1$, …, r, $r+k+1$,…, $r+n$, $\wedge$ I
$r+n+q+3$	$\neg\neg P$	$r+n+1$, $r+n+q+1$, $r+n+q+2$, $\neg$ I
$r+n+q+4$	P	$r+n+q+3$, $\neg$ E

Thus: If $H_1, \ldots, H_m, C_1, \ldots, C_k, \neg P \vdash \neg(H_{m+1} \wedge \ldots \wedge H_r \wedge C_{k+1} \wedge \ldots \wedge C_n)$, then $H_1, \ldots, H_r, C_1, \ldots, C_n \vdash P$.
Theorem 3(d) follows from I and II.

Theorem 3(d) is interesting when the hypotheses $H_1, \ldots, H_m$ are auxiliary hypotheses that are presupposed to falsify the theoretical system $(H_{m+1} \wedge \ldots \wedge H_r \wedge C_{k+1} \wedge \ldots \wedge C_n)$. In this case the auxiliary hypotheses are used as premises in a falsifying argument. Auxiliary hypotheses can be used in this way when they are testable independently of the theoretical system.

It is, however, always possible to regard auxiliary hypotheses as parts of the tested theoretical system. In this case theorems 3(b) and 3(c) are relevant.

There is a third possibility: falsifications presupposing auxiliary hypotheses can be regarded as conditional. In theorem (4) it is shown that if the auxiliary hypotheses $H_1, \ldots, H_m$ are true, then the theoretical system $(H_{m+1} \wedge \ldots \wedge H_r)$ is false.

THEOREM 4:

$H_1, \ldots, H_r, C_1, \ldots, C_n \vdash P$, iff $C_1, \ldots, C_n, \neg P \vdash (H_1 \wedge \ldots \wedge H_m) \rightarrow \neg(H_{m+1} \wedge \ldots \wedge H_r)$ $(0 \leq m \leq r)$

PROOF:

I

Presupposition: $H_1, \ldots, H_r, C_1, \ldots, C_n \vdash P$ (deduction in p steps).

1	C_1	
$\vdots$	$\vdots$	
n	C_n	
$n+1$	$\neg P$	
$n+2$	$\quad H_1 \wedge \ldots \wedge H_m$	
$n+3$	$\quad\quad H_{m+1} \wedge \ldots \wedge H_r$	
$n+4$	$\quad\quad H_1$	$n+2$, $\wedge$ I
$\vdots$	$\quad\quad \vdots$	
$n+m+3$	$\quad\quad H_m$	$n+2$, $\wedge$ I
$n+m+4$	$\quad\quad H_{m+1}$	$n+3$, $\wedge$ I
$\vdots$	$\quad\quad \vdots$	
$n+r+4$	$\quad\quad H_r$	$n+3$, $\wedge$ I
$\vdots$	$\quad\quad \vdots$	
$n+r+p+4$	$\quad\quad P$	$1, \ldots, n, n+4, \ldots, n+r+4$, presupposition
$n+r+p+5$	$\quad\quad \neg P$	$n+1$, rep.
$n+r+p+6$	$\quad \neg(H_{m+1} \wedge \ldots \wedge H_r)$	$n+3, n+r+p+4, n+r+p+5$, $\neg$ I
$n+r+p+7$	$(H_1 \wedge \ldots \wedge H_m) \rightarrow \neg(H_{m+1} \wedge \ldots \wedge H_r)$	$(n+2) - (n+r+p+6)$, $\rightarrow$ I

Thus: if $H_1, \ldots, H_r, C_1, \ldots, C_n \vdash P$, then
$C_1, \ldots, C_n, \neg P \vdash (H_1 \wedge \ldots \wedge H_m) \rightarrow \neg(H_{m+1} \wedge \ldots \wedge H_r)$

II. Presupposition: $C_1, \ldots, C_n, \neg P \vdash (H_1 \wedge \ldots \wedge H_m) \rightarrow \neg(H_{m+1} \wedge \ldots \wedge H_r)$ (deduction in q steps).

1	H_1	
$\vdots$	$\vdots$	
m	H_m	
$m+1$	H_{m+1}	
$\vdots$	$\vdots$	
r	H_r	
$r+1$	C_1	
$\vdots$	$\vdots$	
$r+n$	C_n	
$r+n+1$	$\neg P$	
$\vdots$	$\vdots$	
$r+n+q+1$	$(H_1 \wedge \ldots \wedge H_m) \rightarrow \neg(H_{m+1} \wedge \ldots \wedge H_r)$	$r+1, \ldots, r+n, r+n+1$, presupposition
$r+n+q+2$	$H_1 \wedge \ldots \wedge H_m$	$1, \ldots, m, \wedge$ I
$r+n+q+3$	$\neg(H_{m+1} \wedge \ldots \wedge H_r)$	$r+n+q+1, r+n+q+2, \rightarrow$ E
$r+n+q+4$	$H_{m+1} \wedge \ldots \wedge H_r$	$m+1, \ldots, r, \wedge$ I
$r+n+q+5$	$\neg\neg P$	$r+n+1, r+n+q+3, r+n+q+4, \neg$ I
$r+n+q+6$	P	$r+n+q+5, \neg$ E

Thus: If $H_1, \ldots, H_m, C_1, \ldots, C_k, \neg P \vdash$
$\neg(H_{m+1} \wedge \ldots \wedge H_r \wedge C_{k+1} \wedge \ldots \wedge C_n)$, then $H_1, \ldots, H_r, C_1, \ldots, C_n \vdash P$.
Theorem 4 follows from I and II.

BIBLIOGRAPHY

Agassi, Joseph. 1975. *Science in flux*. Boston studies in the philosophy of science, 28. Dordrecht (Holland): Reidel.

Albert, Hans. 1991. *Traktat über kritische Vernunft*. 5th ed. Tübingen: J. C. B. Mohr (Paul Siebeck).

Andersson, Gunnar. 1981a. Feyerabends Kritik des kritischen Rationalismus. In *Versuchungen: Aufsätze zur Philosophie Paul Feyerabends*, ed. Hans Peter Duerr, 2: 159-74. Frankfurt am Main: Suhrkamp.

——. 1981b. Sind Falsifikationismus und Fallibilismus vereinbar? In *Voraussetzungen und Grenzen der Wissenschaft*, ed. Gerard Radnitzky, and Gunnar Andersson, 255-76. Die Einheit der Gesellschaftswissenschaften 25. Tübingen: J. C. B. Mohr (Paul Siebeck).

——. 1982. Naive and critical falsificationism. In *In pursuit of truth: Essays on the philosophy of Karl Popper on the occasion of his 80th birthday*, ed. Paul Levinson, 50-63. Atlantic Highlands (New Jersey); Brighton (Sussex): Humanities Press; Harvester Press.

——. 1984. How to accept fallible test statements? Popper's criticist solution. In *Rationality in science and politics*, ed. Gunnar Andersson, 47-68. Boston Studies in the Philosophy of Science 79. Dordrecht (Holland): Reidel.

——. 1985. Basic statements and incommensurability. *Methodology and Science: Interdisciplinary Journal for the Empirical Study of the Foundations of Science and Their Methodology*, 18 (3): 186-98.

——. 1991a. Feyerabend on falsifications, Galileo, and Lady Reason. In *Beyond reason: Essays on the philosophy of Paul Feyerabend*, ed. Gonzalo Munévar, 281-95. Boston Studies in the Philosophy of Science, 132. Dordrecht (Holland): Kluwer Academic Publishers.

——. 1991b. The tower experiment and the Copernican revolution. *International Studies in the Philosophy of Science*, 5 (2): 143-52.

Aston, F. W. 1938. Forty years of atomic theory. In *Background to modern science: Ten lectures at Cambridge arranged by the history of science committee 1936*, ed. Josph Needham, and Walter Pagel, 93-114. New York: Macmillan; reprint, New York: Arno Press, 1975.

Bacon, Francis. 1620. *The new organon and related writings*. F. H. Anderson ed. New York: Liberal Arts Press, 1960.

Bartley, William Warren, III. 1973. *Wittgenstein*. Philadelphia (Pennsylvania): J. B. Lippincott.

——. 1982. The philosophy of Karl Popper, Part III: Rationality, criticism, and logic. *Philosophia: Philosophical Quarterly of Israel*, 11 (1-2): 121-221.

——. 1984. *The retreat to commitment*. 2nd, revised and enlarged ed. La Salle (Illinois): Open Court Publishing Company.

Bernard, Claude. 1957. *An introduction to the study of experimental medicine*. New York: Dover Publications.

Berry, Arthur. 1961. *A short history of astronomy: From earliest times though the nineteenth century*. Reprint, New York: Dover.

Campbell, Donald T. 1974. Evolutionary epistemology. In *The philosophy of Karl Popper*, ed. Paul Arthur Schilpp, 1: 413-63. The Library of Living Philosophers. La Salle (Illinois): Open Court.

Conant, James Bryant. 1957. The overthrow of the phlogiston theory: The chemical revolution of 1775-1789. In *Harvard case histories in experimental science*, ed. James Bryant Conant, and Leonard K. Nash, 2: 65-115. Cambridge (Mass.): Harvard University Press.

Copernicus, Nicholas. 1939. *Commentariolus*. In *Three Copernican treatises*, ed. Edward Rosen, 55-89. Reprint, New York: Octagon Books, 1971.

Copi, Irving M. 1986. *Introduction to logic*. 7th ed. New York: Macmillan Publishing Company.

Craig, Robert P. 1982. Loch Naess: The monster unveiled. *New Scientist*, 95: 354-57.

Descartes, René. 1637. *Discours de la méthode*.

Dijksterhuis, E[chard] J. 1961. *The mechanization of the world picture*. London: Oxford University Press.

Drake, Stillman. 1970. *Galileo studies: Personality, tradition, and revolution*. Ann Arbor (Michigan): The University of Michigan Press.

Dreyer, J. L. E. 1953. *A history of astronomy from Thales to Kepler*. 2d ed. Reprint of edition first published 1906 under the title "History of the planetary systems from Thales to Kepler"; New York: Dover.

Duhem, Pierre. 1974. *The aim and structure of physical theory*. New York: Atheneum.

Feigl, Herbert. 1971. Reserach programmes and induction. In *PSA 1970: In memory of Rudolf Carnap*, ed. Robert C. Buck, and Roger S. Cohen, 147-50. Proceedings of the 1970 biennial meeting, philosophy of science association. Boston studies in the philosophy of science, 8. Dordrecht (Holland): Reidel.

Feyerabend, Paul K. 1970. Consolations for the specialist. In *Criticism and the growth of knowledge*, ed. Imre Lakatos, and Alan Musgrave, 197-230. London: Cambridge University Press.

——. 1975. *Against method: Outline of an anarchistic theory of knowledge*. London: NLB.

——. 1976. On the critique of scientific reason. In *Essays in memory of Imre Lakatos*, ed. Robert S. Cohen, Paul K. Feyerabend, and Max W. Wartofsky, 109-43. Boston studies in the philosophy of science, 39. Dordrecht (Holland): Reidel.

——. 1978. *Der wissenschaftstheoretische Realismus und die Autorität der Wissenschaften*. Vol. 1 of *Ausgewählte Schriften*. Braunschweig: Vieweg.

——. 1981a. *Philosophical papers*. 2 vols. Cambridge: Cambridge University Press.

——. 1981b. Rückblick. In *Versuchungen: Aufsätze zur Philosophie Paul Feyerabends*, ed. Hans Peter Duerr, 2: 320-72. Frankfurt am Main: Suhrkamp.

——. 1983. *Wider den Methodenzwang*. 2 ed. Frankfurt am Main: Suhrkamp.

——. 1988. *Against method*. Rev. ed. London: Verso.

Fleck, Ludwik. 1981. *Genesis and development of a scientific fact*. Thaddeus J. Trenn, and Robert K. Merton, ed. Chicago: University of Chicago Press.

—— 1983. *Erfahrung und Tatsache: Gesammelte Aufsätze*. Frankfurt am Main: Suhrkamp.

Galilei, Galileo. 1957. *The starry messenger*. In *Discoveries and opinions of Galileo*, 21-58. Garden City, NY: Anchor-Doubleday.

——. 1967. *Dialogue concerning the two chief world systems - Ptolemaic & Copernican*. 2 ed. Berkeley: University of California Press.

Gregory, R. L. 1973. The confounded eye. In *Illusion in nature and art*, ed. R. L. Gregory, and E. H. Gombrich, 45-95. London: Duckworth.

Grize, Jean-Blaise. 1972. *Logique des propositions et des prédicats, déduction naturelle*. Vol. 1 of *Logique moderne*. 2 ed. Mathématiques et sciences de l'homme, 10. Paris: Mouton/Gauthier-Villars.

Grosser, Morton. 1979. *The discovery of Neptune*. New York: Dover.

Grünbaum, Adolf. 1960. The Duhemian argument. *Philosophy of Science*, 27: 75-87.

Hanson, Norwood Russell. 1958. *Patterns of discovery: An inquiry into the conceptual foundations of science*. London: Cambridge University Press.

Hanson, Norwood Russell. 1962. Leverrier: The zenith and nadir of Newtonian mechanics. *Isis*, 53: 359-78.

Hempel, Carl G. 1965. *Aspects of scientific explanations and other essays in the philosophy of science*. New York and London: Free Press and Collier-Macmillan.

——. 1966. *Philosophy of natural science*. Prentice-Hall Foundations of Philosophy Series. Englewood Cliffs (N. J.): Prentice-Hall, Inc.

Hume, David. 1739–40. *A treatise of human nature*. Ed. L. A. Selby-Bigge. 2d ed. Oxford: Clarendon Press, 1967.

Hübner, Kurt. 1979. *Kritik der wissenschaftlichen Vernunft*. 2 ed. Freiburg: Verlag Karl Alber.

Kleene, Stephen Cole. 1967. *Mathematical logic*. New York: John Wiley & Sons.

Kneale, William C. 1974. The demaraction of science. In *The philosophy of Karl Popper*, ed. Paul Arthur Schilpp, 1: 205-217. The Library of Living Philosophers. La Salle (Illinois): Open Court.

Koertge, Noretta. 1978. Towards a new theory of scientific inquiry. In *Progress and rationality in science*, ed. Gerard Radnitzky, and Gunnar Andersson, 253-78. Boston studies in the philosophy of science, 58. Dordrecht (Holland): Reidel.

Koyré, Alexandre. 1957. *From the closed world to the infinite universe*. Baltimore: Johns Hopkins University Press.

Kuhn, Thomas S. 1957. *The copernican revolution: Planetary astronomy in the development of western thought*. Cambridge (Mass.): Harvard University Press.

——. 1970a. Logic of discovery or psychology of research? In *Criticism and the growth of knowledge*, ed. Imre Lakatos, and Alan Musgrave, 1-23. London: Cambridge University Press.

——. 1970b. *The structure of scientific revolutions*. 2nd, rev. ed. Chicago: University of Chicago Press.

Lakatos, Imre. 1968a. Changes in the problem of inductive logic. In *The problem of inductive logic*, ed. Imre Lakatos, 315-417. Amsterdam: North-Holland.

——. 1968b. Criticism and the methodology of scientific research programmes. *Proceeding of the Aristotelian Society*, 69: 149-86.

——. 1970. Falsification and the methodology of scientific research programmes. In *Criticism and the growth of knowledge*, ed. Imre Lakatos, and Alan Musgrave, 91-195. London: Cambridge University Press.

——. 1971a. History of science and its rational reconstructions. In *PSA 1970: In memory of Rudolf Carnap*, ed. Robert C. Buck, and Roger S. Cohen, 91-135. Proceedings of the 1970 biennial meeting, philosophy of science association. Boston studies in the philosophy of science, 8. Dordrecht (Holland): Reidel.

——. 1971b. Replies to critics. In *PSA 1970: In memory of Rudolf Carnap*, ed. Robert C. Buck, and Roger S. Cohen, 174-81. Proceedings of the 1970 biennial meeting, philosophy of science association. Boston studies in the philosophy of science, 8. Dordrecht (Holland): Reidel.

——. 1974. Popper on demarcation and induction. In *The philosophy of Karl Popper*, ed. Paul Arthur Schilpp, 1: 241-73. The Library of Living Philosophers. La Salle (Illinois): Open Court.

——. 1978. *Philosophical papers*. 2 vols. Cambridge: Cambridge University Press.

——, and Zahar, Elie. 1975. Why did Copernicus' research program supersede Ptolemy's? In *The Copernican achievement*, ed. Robert S. Westman, 354-83. Contributions - UCLA center for medieval and renaissance studies, 7. Berkeley: University of California Press.

Le Roy, Édouard. 1899–1900. Science et philosophie. *Revué de métaphysique et de morale*, 7-8: 7: 375-425, 503-62, 706-31; 8: 37-72.

Musgrave, Alan. 1976. Method or madness? In *Essays in memory of Imre Lakatos*, ed. Robert S. Cohen, Paul K. Feyerabend, and Max W. Wartofsky, 457-91. Boston studies in the philosophy of science, 39. Dordrecht (Holland): Reidel.

Poincaré, Henri. 1970. *La valeur de la science.* [Paris]: Flammarion.

Popper, Karl R. 1959. *The logic of scientific discovery*. Revised ed. London: Hutchinson, 1972.

——. 1960. *The poverty of historicism.* 2 ed. London: Routledge & Kegan Paul.

——. 1965. *Conjectures and refutations: The growth of scientific knowledge.* 2 ed. London: Routledge and Kegan Paul.

——. 1966. *The open society and its enemies*. 2 vols. 5 ed. London: Routlege and Kegan Paul.

——. 1970a. Die Logik der Sozialwissenschaften. In *Der Positivismusstreit in der deutschen Soziologie*, ed. Heinz Maus, and Friedrich Fürstenberg, 103-123. Neuwied: Hermann Luchterhand Verlag GmbH.

——. 1970b. Normal science and its dangers. In *Criticism and the growth of knowledge*, ed. Imre Lakatos, and Alan Musgrave, 51-58. London: Cambridge University Press.

——. 1972. *Objective knowledge: An evolutionary approach*. London: Oxford University Press.

——. 1973. *Logik der Forschung*. 5 ed. Tübingen: J. C. B. Mohr (Paul Siebeck).

——. 1974. Replies to my critics. In *The philosophy of Karl Popper*, ed. Paul Arthur Schilpp, 2: 961-1197. La Salle (Illinois): Open Court.

——. 1976. The myth of the framework. In *The abdication of philosophy: Philosophy and the public good. Essays in honor of Paul Arthur Schilpp*, ed. Eugene Freeman, 23-48. La Salle (Illinois): Open Court?

——. 1979. *Die beiden Grundprobleme der Erkenntnistheorie*. Tübingen: J. C. B. Mohr (Paul Siebeck).

——. 1983. *Realism and the aim of science; From the postscript to the Logic of scientific discovery.* Totowa (New Jersey): Rowman and Littlefield.

Price, Derek J. de Solla. 1959. Contra-Copernicus: A critical re-estimation of the mathematical planetary theory of Ptolemy, Copernicus, and Kepler. In *Critical problems in the history of science*, ed. Marshall Clagett, 197-218. Madison: University of Wisconsin Press.

Pähler, Klaus. 1979. Kritik ohne Erkenntnisfortschritt? Probleme Popperianischer Wissenschaftstheorie. Ph.D. diss., Trier University.

Quine, Williard Van Orman. 1961. *From a logical point of view: 9 logico-philosophical essays.* 2d, rev. ed. Cambridge (Mass.): Harvard Univ. Press.

——. 1991. Two dogmas in retrospect. *Canadian Journal of Philosophy*, 21 (3, September): 265-74.

Radnitzky, Gerard. 1973. *Contemporary schools of metascience*. 3 ed. Chicago: Regnery.

Ronchi, Vasco. 1966. *L'optique: Science de la vision*. Paris: Masson.

Russell, Bertrand. 1940. *An inquiry into meaning and truth.* London: George Allen and Unwin.

——. 1961. *History of western philosophy and its connection with political and social circumstances from the earliest times to the present day*. 2d ed. London: George Allen & Unwin Ltd.

Schäfer, Lothar. 1974. *Erfahrung und Konvention: Zur Theorienbegriff der empirischen Wissenschaften*. Stuttgart: Friedrich Frommann Verlag.

Stegmüller, Wolfgang. 1971. Das Problem der Induktion: Humes Herausforderung und moderne Antworten. In *Neue Aspekte der Wissenschaftstheorie*, ed. Hans Lenk, 13–74. Braunschweig:

——. 1973. *Probleme und Resultate der Wissenschaftstheorie und der Analytischen Philosophie*. Vol. 2, *Theorie und Erfahrung*. Part 2, *Theorienstrukturen und Theoriendynamik*. Berlin: Springer.

——. 1974. *Probleme und Resultate der Wissenschaftstheorie und der Analytischen Philosophie*. Vol. 1, *Wissenschaftliche Erklärung und Begründung*. Berlin: Springer.

Stegmüller, Wolfgang. 1979. *Hauptströmungen der Gegenwartsphilosophie: Eine kritische Einführung*. 4 vols. Vol. 2. 6 ed. Stuttgart: Kröner.
Stimson, Dorothy. 1917. *The gradual acceptence of the Copernican theory of the universe*. Hanover, N. H.:
Ströker, Elisabeth. 1976. *Wissenschaftsgeschichte als Herausforderung: Marginalien zur jüngsten wissenschaftstheoretischen Kontroverse*. Wissenschaft und Gegenwart: Geisteswissenschaftliche Reihe, 59/60. Frankfurt am Main: Vittorio Klostermann.
Van Helden, Albert. 1974. The telescope in the seventeenth century. *Isis*, 65: 38-58.
Watkins, John. 1984. *Science and scepticism*. London: Hutchinson.
Wittgenstein, Ludwig. 1967. *Philosophical investigations*. 3 ed. Oxford: Basil Blackwell.
Wolf, Rudolf. 1890–92. *Handbuch der Astronomie, ihrer Geschichte und Litteratur*. 2 vols. Zürich: F. Schulthess.
Worrall, John. 1978. The ways in which the methodology of scientific research programmes improves on Popper's methodology. In *Progress and rationality in science*, ed. Gerard Radnitzky, and Gunnar Andersson, 45-70. Boston studies in the philosophy of science, 58. Dordrecht (Holland): Reidel.

INDEX OF SUBJECTS

INDEX OF NAMES

Philosophy of History and Culture

This series presents original books broadly concerned with philosophical treatments of the ideas of history and culture, with historically and culturally embodied entities, and with methodologies and interpretive strategies pertinent to their explanation and understanding.

1. HERTZBERG, L. and J. PIETARINEN (eds.). *Perspectives on Human Conduct.* 1988. ISBN 90 04 08937 3
2. DRAY, W.H. *On History and Philosophers of History.* 1989. 90 04 09000 2
3. ROTENSTREICH, N. *Alienation.* The Concept and its Reception. 1989. 90 04 09001 0
4. ORUKA, H.O. *Sage Philosophy.* Indigenous Thinkers and Modern Debate on African Philosophy. 1990. ISBN 90 04 09283 8
5. MERCER, R. *Deep Words.* Miura Baien's System of Natural Philosophy. 1991. ISBN 90 04 09351 6
6. VAN DER DUSSEN, W.J. and L. RUBINOFF (eds.). *Objectivity, Method and Point of View.* 1991. ISBN 90 04 09411 3
7. DASCAL, M. (ed.). *Cultural Relativism and Philosophy.* North and Latin American Perspectives. 1991. ISBN 90 04 09433 4
8. WHITE, F.C. *On Schopenhauer's* Fourfold Root of the Principle of Sufficient Reason. 1992. ISBN 90 04 09543 8
9. ZEMACH, E.M. *Types.* Essays in Metaphysics. 1992. ISBN 90 04 09500 4
10. FLEISCHACKER, S. *Integrity and Moral Relativism.* 1992. ISBN 90 04 09526 8
11. VON WRIGHT, G.H. *The Tree of Knowledge and Other Essays.* 1993. ISBN 90 04 09764 3
13. ANDERSSON, G. *Criticism and the History of Science.* Kuhn's, Lakatos's and Feyerabend's Criticisms of Critical Rationalism. 1994. ISBN 90 04 10050 4
14. VADEN HOUSE, D. *Without God or His Doubles.* Realism, Relativism and Rorty. 1994. ISBN 90 04 10062 8

E.J. Brill — P.O. Box 9000 — 2300 PA Leiden — The Netherlands